Rescuing Wild Animals
and Living to Tell About It

Bob Anderson

Krittergitters Publishing
Wills Point, Texas

Rescuing Wild Animals and Living to Tell About It

The contents of this work, including, but not limited to, the accuracy of events, people, and places depicted; opinions expressed; permission to use previously published materials included; and any advice given or actions advocated are solely the responsibility of the author, who assumes all liability for said work and indemnifies the publisher against any claims stemming from publication of the work.

The information in this book is true and complete to the best of our knowledge. The author and the publisher disclaim any liability in connection with the use of this information.

ISBN: 978-164008495-7

Printed in the United States of America

Second Printing

Book design by Lee Lewis Walsh, Words Plus Design, www.wordsplusdesign.com

For more information or to order additional books, please contact:
Krittergitters Publishing
320 VZCR 3219
Wills Point, Texas 75169
U.S.A.
1-903-865-1013
www.krittergitters.com

Contents

Introduction

Anyone who really loves animals has, at some point in their life, probably wished they could do more to help animals in distress. It's very difficult to watch an animal suffer and not be able to help. Fortunately there are people who have learned how to help. These people usually refer to themselves as "animal rehabilitators."

My wife and I made the commitment to rehabilitate wild animals because we wanted to make a difference. We started our rehab career by contacting a licensed animal care facility and became volunteers. It didn't take long to discover that caring for wild animals is not easy. We were also interested in going out into the field and rescuing wild animals in need. We soon discovered this was also a complicated task that required more knowledge than we possessed.

As the years passed, we found ourselves on a very steep learning curve. We continued to become more proficient at rehabilitation and rescue. We also discovered the more we learned, the more we found we were only scratching the surface of valuable information.

After twenty years of working in this field, we determined the field rescue was an area that was supported by very few organizations. Rehab centers are more plentiful and easier to locate, but most don't specialize in field rescue. To help fill this gap, we opened a training academy to teach volunteers to safely rescue, stabilize, and transport critters in need.

To be successful at rehab or rescue requires a good understanding of animals. We found that volunteers who take the time to become experts on particular species are much more successful in rescue and rehab than those who are not experts.

Looking back, we wish there had been a book of basics to keep us moving in the right direction. This book is being prepared as a basic text of what we wish we had had twenty years ago. We have also included some technical information about many of the animals we rescue, some rescue tips from our academy, and some ideas on coexisting with wildlife.

Keep in mind while reviewing this information, there is always more than one way to complete any task. Gathering information from many sources and coupling the information with your own experiences is the best way to become successful.

Helping Animals

Wild animals have been roaming the earth for millions of years. As the world has changed over time, wild animal species have come and gone. Natural disasters, changing climates, and environmental changes have had major effects on the wild creatures.

One could argue that the greatest long-term negative impacts on animals began when humans came upon the scene. The animals provided almost everything mankind needed to survive. Mankind, on the other hand, gave the animals very little. This one-sided relationship has continued throughout the ages. A reasonable balance was maintained primarily because both man and animal seemed to propagate at reasonably constant levels. As mankind started rapidly expanding into areas once reserved for animal habitat, the balance began to shift. As this shift progressed, animal and human conflicts started becoming more common. Animals that at one time were considered valuable natural resources suddenly found themselves viewed as pests. As the animal habitats continued to decrease, the little creatures started to rely on humans as a means for their survival. Suddenly people no longer needed the animals and wanted them out of the way.

As the threat to wildlife increases, community efforts to provide humane treatment and care of wildlife also increase.

This book provides information to assist interested parties to become more proficient in helping our dwindling wildlife populations. We discuss procedures, tips, animal data, and some technical data needed to successfully rescue a wild creature in need. This book will not cover specifics on rehabilitation but instead will focus more on handling and rescue.

Those dedicated to helping animals obviously make a difference but do not hold the key to solving the increasing problem of animal conflict. The real solution is learning to coexist with our wild neighbors. The first step to making this happen is education. People need to understand the animals, their habits, their needs, and what they can do to live in harmony. This concept will be discussed in greater detail in a later chapter.

There are several areas of expertise the animal helper should posses. Although each species has very unique requirements for foods, medicines, health care, and stress management, many of the basic protocols are the same for any animal. The following information is universal and basic to the care of any rescued animal.

Safety

The single most important part of working with wild animals is being safe. When handling animals, there are two rules that can never be violated.

The first rule is: If you are not sure you can complete the task without being injured, do not attempt the task!

The second rule is: If you are not sure you can complete the task without being injured, do not attempt the task!

Almost all animals have one thing in common: They will bite to protect themselves, and they are totally unpredictable. The ones that don't bite have other ways of fending off predators: claws, tails, spray, or kicking. Whatever the mechanism, the animal can hurt you. Animal wounds are always serious and often result in long-term concern. Before attempting to handle a wild animal, take the time to learn how the particular species protects itself. Find out if the species is normally aggressive and determine what zoonotic diseases are common for that animal. Most states require that any wild animal that bites a person must be destroyed. If you are careless and allow yourself to be injured, you have actually made a bad situation worse.

Going into the field to pick up an injured or sick wild animal can be a little tricky. There are many things that come into play when going after an angry animal. The situation can be dangerous to both the rescuer and the animal. Every rescue will be unique and will require careful execution, but following set procedures can increase the probability of a successful rescue. If you get in the habit of starting your rescue in a way that has proven successful and then are consistent in rescues, you will make fewer mistakes.

Chapter One

Zoonosis

This book covers a variety of topics pertaining to rescue, first-aid, and coexisting with wildlife. Before anyone attempts to interact with wild animals, they should be aware of any hazards associated with these activities. The most obvious threat comes from injury caused by the animal biting and scratching. All wild animals will bite when frightened or threatened. In addition to physical injuries, animal caretakers are also at high risk to contract zoonotic diseases. Zoonosis is a term that is used to identify any disease transported from vertebrate animals to humans. Wild animals can serve as primary or secondary sources of zoonoses. That is, transmission of zoonotic diseases can occur by direct means (via contact with the animal, its urine, or its feces) or indirectly (by air, soil, and water contaminated with the infectious agent). Vectors, such as ticks and mosquitoes, also play an important role in the transmission of diseases.

We feel it is important to be aware of some of the most common zoonotic diseases.

Rabies (Lyssavirus)

Any warm-blooded mammal can carry rabies, but the most common carriers are striped skunks, raccoons, bats, foxes, and coyotes.

There are seven distinct variants or strains of rabies that affect land mammals as well as an unknown number that affect bats. The various strains appear dominant in different geographic regions. Since 1960 rabies in the United States has been more frequently reported in wild than in domestic animals. Wildlife now accounts for more than 90 percent of all rabies cases.

Raccoons, skunks, foxes, coyotes, and bats account for the majority of cases in the United States. Coyote rabies has reached epizootic (attacks many animals in a region at the same time) proportions in southern Texas. Bats comprise the third-largest rabies reservoir after skunks and raccoons. It is reported, however, that less than 1 percent of the bat population carry the disease.

Human rabies continues to be reported in small numbers: Twenty-four cases were reported from 1980 to 1994. Fourteen individuals were diagnosed with strains known from animal reservoirs in the U.S., and ten were traced to strains associated with bats.

Most cases of rabies are the result of a bite from an infected animal. A non-bite exposure can also occur when saliva or brain tissue from an infected animal enters open skin or mucous membranes (eyes, nose, or mouth).

Humans are relatively resistant to rabies. Reports indicate only 15 percent of humans bitten by a known rabid animal become infected with rabies.

Symptoms of Rabies in Humans

Reliable incubation periods range from fewer than ten days to more than six years. After initiation of clinical signs, the virus is almost always fatal.

Early signs are characterized by apprehension, headache, a low-grade fever accompanied by vague sensory changes such as discomfort and irritation or pain in the area of the bite, and swelling of the lymph nodes. Visual and auditory sensitivities may occur, causing disorientation, agitation, hallucination, and (rarely) aggression.

Difficulty swallowing due to muscle spasms and hypersalivation lead to hydrophobia (foaming at the mouth) in approximately 50 percent of the cases. Partial paralysis and, eventually, complete paralysis ensues. Death usually results from respiratory failure one week after onset of the disease.

Prevention and Treatment

1. Always wear heavy gloves when handling caged animals.
2. Be extremely cautious of trapped animals acting strangely. If the animal appears sick or injured, isolate it and consult a veterinarian or licensed rehabilitator.
3. Do not pick up animals not in a cage. Rescue of sick and injured animals requires two people.
4. Don't let your face get close to the trap. Animals scratch.
5. If you have doubts about the situation, ask for help.

Any wild animal bite requires immediate medical attention.

If you experience a bite from any wild animal that breaks the skin, there are several things that must be done:

1. Do not release the animal if at all possible, but do not put yourself in a position for further injuries. In accordance with most state laws, any wild animal that bites a person must be destroyed. It is also crucial to have the animal available for rabies testing. If the animal is not available for testing, you will likely be given the post-exposure preventative rabies inoculation series. This series of injections has been greatly modified from the former procedures (multiple injections in the abdomen) to one in which seven to ten shots are given at the wound site and in the upper arm.
2. Scrub any bite wound immediately and aggressively with soap and water. If an antiseptic soap such as Betadine is available, use that. Flush thoroughly with water to facilitate organism removal.
3. You should get immediate medical attention.

Hantavirus (Sin Nombre Virus)

In 1993 seventeen people in the Four-Corners area (that area where Utah, Colorado, New Mexico, and Arizona share a common point) died within a period of a few weeks from extreme respiratory failure caused by the lungs filling with fluids. This cluster of deaths triggered a major search for the cause. In May of 1993, the Hantavirus was identified and confirmed. The disease was officially termed Hantavirus Pulmonary Syndrome (HPS). To date, there have been eight species of the Hantavirus identified in the United States. By 1997 one hundred fifty-eight cases had been verified from twenty-seven different states.

The primary carrier of the Hantavirus is the deer mouse. It has, however, been found in rice rats and cotton rats. Hantaviruses are rodent-borne hemorrhagic fever viruses and have been known as important agents of human disease throughout Asia and Europe for hundreds of years.

Hantavirus is maintained in the deer mouse as a persistent infection characterized by long-term shedding in the urine, feces, or saliva. Humans become infected by inhalation of small particle aerosols (dust) of virus-contaminated feces, urine, or saliva. Rodent bites could be another avenue of virus transmission.

Symptoms of Hantavirus in Humans

After the initial exposure, Hantavirus undertakes a two-to four-week incubation period before the disease ensues. Previously healthy individuals suddenly develop symptoms of headache, fever, pain in muscles, dry cough, dizziness, and intestinal problems, including nausea, vomiting, and diarrhea. The first symptoms mimic a typical upper respiratory virus. These flu-like symptoms progress to acute pulmonary (lungs) insufficiency, indistinguishable from adult respiratory distress syndrome. Within days of onset, death from respiratory failure can occur.

Prevention and Treatment

When working in places infested with rodents, and specifically when trapping small rodents, protective clothing, including heavy protective gloves, should be worn.

Work should be done in a well-ventilated place whenever possible. A respirator fitted with a high-efficiency particulate filter is recommended. At a minimum, a good-quality dust mask should be used. It is reported that the Hantavirus is a large virus (>100 microns in diameter) and can be filtered with a standard dusk mask.

Another precaution to prevent exposure is to disinfect old areas of contact with a 1:10 solution of household bleach and water.

An effective treatment for this virus is still being researched. Mortality rates exceed 50 percent, despite aggressive treatment.

Rocky Mountain Spotted Fever (Rickettsia Ricketts)

Rocky Mountain Spotted Fever (tick fever) is the second-most common tick-borne disease in the United States and is caused when a human is infected by rickettsias. Rickettsias are tiny organisms between viruses and bacteria in size. Rickettsias are unable to reproduce unless attached to a host. The disease is transmitted from a reservoir rodent to a human via ticks. The tick bites the rodent, acquiring the rickettsias, then bites the human, transmitting into the bloodstream. Carried on rabbits, rodents, and the Virginia opossum, the American dog tick in the East and the American wood tick in the West are the regional primary vectors.

Symptoms of Rocky Mountain Spotted Fever in Humans

After a human is bitten, there is a delay in the actual transmission of the Rickettsia (the disease-causing microorganism) from the tick to the host. A reactivation process occurs, leading to a four-to twenty-hour delay of infection. Human infection is characterized by a three-to fourteen-day incubation period followed by non-specific signs of the disease such as lethargy and weakness, rapidly progressing to fever, chills, headache, muscle pain, and pain in the joints. Fever is the most predictable and persistent sign. In 88 percent of the cases, generalized spotting or blotchiness starts first on the extremities, next on the palms and soles, and then rapidly spreads to the trunk. Patchily (pinpoint) hemorrhages are common.

In the absence of a diagnosis and specific therapy, a 15 to 20 percent mortality rate results. Even with effective anti-microbial therapy, a 2 percent mortality rate in the U.S. results. Annually, about fifty people die from this disease (six hundred to one thousand two hundred cases are reported each year). Early diagnosis and initiation of therapy, before serologic or isolation evidence, leads to a rapid uneventful recovery in most cases.

Prevention

Procedures to avoid contact with ticks are the primary recommendations. Wearing gloves, long pants, and long-sleeved shirts, along with conducting frequent checks of clothing and body, are recommended. Remove ticks without crushing them and thoroughly scrub the wound.

Encephalitis

Encephalitis is an inflammation of the brain caused by one of several togaviruses. These viruses are transmitted to humans by mosquitoes and ticks. The most common hosts are birds, but the virus has also been found in rabbits and rodents. There are five common varieties of encephalitis in the United States. In humans some varieties of the virus result in minor symptoms, sometimes not even requiring treatment. Others, however, can be very severe, even fatal. Outbreaks usually occur in the spring and summer following rainy seasons with high mosquito populations.

Symptoms of Encephalitis in Humans

Incubation is usually four to six days after being bitten by an infected mosquito or tick. Early symptoms include headaches, high fever, and pain and weakness in the muscles. These symptoms have a tendency to go away after several days and then reappear again.

Raccoon Roundworm—Baylisascaris Procyonis (Larval Migrans)

Found primarily in raccoons, larval migrans is a disease process started by the ingestion of the eggs of the raccoon roundworm, Baylisascaris Procyonis. It involves the prolonged migration and persistence of the larval parasite in internal organs and tissues.

Baylisascaris adults reside in the small intestine of raccoons, causing disease primarily in heavily parasitized young raccoons. The adult worms produce and shed eggs in the feces. Adults are known to shed up to six million eggs per day. These become infective in three to four weeks under most environmental conditions. In the Midwest, studies indicate that 70 to 85 percent of the raccoons are infected.

Due to high population densities of raccoons and the persistence of eggs to live for months to years, the potential for human exposure is high. Animal handlers are at high risk for exposure.

Symptoms of Baylisascaris Procyonis in Humans

The most common manifestations of Baylisascaris are minor tissue damage and ultimate encapsulation of the parasite in non-critical sites such as skeletal muscle. Large numbers of parasites, however, can lead to more serious symptoms, including fever, blood disorders, and lung conditions similar to pneumonia. Most seriously, central nervous system (CNS) disease is possible. Baylisascaris can likely develop in the tissues of the nervous system, in particular, the brain. CNS disease depends on the number of larvae entering the brain, the location, extent of migration, and size of the brain. Progressive CNS disease has resulted in the deaths of at least two children who ingested large numbers of larvae. Ocular larva migrans results when larvae are delivered to the eye by the systemic circulation. Migration of larvae in the retina can cause loss of vision or acute sensitivity to light.

Prevention

The key to preventing exposure is to reduce contact with feces or areas where fecal matter has lain. The eggs are highly resistant to environmental and chemical disinfectants and can remain adhering to surfaces and dormant in soil for long periods of time.

Cages, traps, and rehabilitation areas are sources of contamination and should be handled cautiously. Gloves should always be worn when cleaning or working around contamination sources. Traps and cages should be regularly cleaned and disinfected using a 1:10 mixture of Clorox™ and water. While not destroying the eggs, this will help remove the protein coat, which makes them sticky enough to adhere to most surfaces. The only accepted method of killing the eggs is through the use of a blowtorch. Propane weed burners work well. When cleaning with a power washer, a face shield should be used to prevent blow-back contamination.

Bubonic Plague (Yersina Pestis)

Plague is a disease caused by bacterium, Yersina pestis, which is transmitted via fleabites. People can also contract the disease by coming into contact with infected animal flesh. Plague is one of the most dangerous zoonotic diseases. There are three forms of plague found in humans: bubonic, septicemic, and pneumonic, all caused by the same

bacterium. Plague has over two hundred known carriers. Plague cases are very rarely reported in most parts of the United States. Ninety percent of the reported cases are found in New Mexico, Arizona, California, and Colorado. When rodent numbers are great and are accompanied by dense flea populations, periodic epizootics occur. These result in high mortality to the rodents, which then increases the possibility of exposure to domestic pets and humans when the fleas look for alternate hosts. Bubonic plague is primarily a flea-transmitted disease but can be acquired through direct contact with infected animals and their tissues or by inhalation.

Symptoms of Bubonic Plague in Humans

After a two- to six-day incubation period, an infection emanates from the site of the fleabite. In bubonic plague, characteristic symptoms include fever, headache, chills, muscle pain, weakness, fatigue, and upset stomach. In bubonic plague, lymph nodes enlarge, forming "buboes" (enlarged and inflamed lymphs) especially in the region of first exposure. Septicemic plague is a secondary-stage bubonic plague infecting the bloodstream. Those infected with septicemic plague do not present with fever or buboes. In pneumonic plague, the bacteria settles in the lungs. This form of plague can be transmitted between humans. Case fatality rates reach 50 to 60 percent in untreated cases of bubonic plague and nearly 100 percent in pneumonic cases. Death usually occurs within days.

Symptoms of Bubonic Plague in Rodents

Animals that are infected by the disease will become very lethargic and inactive. They can easily be approached and will usually die shortly after the disease takes hold. Rodent die-offs may signal the start of a period of special concern and increased precautions.

Prevention and Treatment

With early detection, bubonic plague can be successfully treated with antibiotics. The best method of prevention is to control your exposure to fleas. When handling and relocating rodents, gloves and a long-sleeved shirt should be worn and each animal should be dusted with flea powder. Besides reducing the exposure to workers, this also reduces the spread of the disease through relocation. If you develop a high fever, be sure to tell your physician that you are high risk for plague.

Tularemia (Francisella Tularensis)

Tularemia (rabbit fever) is caused by the bacterium Francisella Tularensis. It is carried primarily by rabbits but has also been found in approximately one hundred other rodent and rabbit species. It is transmitted through the bite of flies, fleas, ticks, or lice. There are two types of the disease: Type A, which is the more virulent form carried by the vector species, and Type B, a milder form that can be carried by water. There are approximately two hundred cases of tularemia reported in humans each year, but under-reporting is suspected. Biting flies, ticks, and direct contact with the infected animal are equally responsible as sources of infection. Water-borne tularemia predominates in northern regions where muskrats and beavers contaminate surface waters.

Symptoms of Tularemia in Humans

After an incubation time of two to ten days, tularemia can take at least four different forms, depending on the method of infection. The most common clinical course is the ulcero-glandular form. At the site of the entry, usually the hands or at the site of an arthropod bite, a necrotic ulcer develops and the organism spreads to the regional lymph node. Symptoms include chills, high fever, severe headaches, muscle aches, joint pain, nausea, and fatigue. When the organism contaminates the eye, an ulcer often forms on the lower lid. Other forms of the disease include glandular "of the glands," without a skin lesion), typhoidal (post-ingestion), and oropharyngeal. The case fatality rate in untreated cases is 5 to 7 percent.

Prevention and Treatment

As with other diseases transmitted by tick or insect vectors, handling should be kept to a minimum and gloves should always be worn. You should always practice good hygiene. It is suggested that any trapped rabbit be dusted with flea and tick powder prior to being handled.

Leptospirosis (Leptospira Interrogans)

Leptospirosis is caused by the bacterium Leptospira interrogans and is transmitted via urine from infected animals. Raccoons, opossums, Norway rats, mice, white-tailed deer, and striped skunks carry this disease. Striped skunks are a significant reservoir with some populations demonstrating a greater-than-50 percent prevalence. Skunks can shed

leptospires in their urine for up to three hundred days. Transmission occurs by direct contamination of mucus membranes or broken skin by urine. Indirectly, it can be acquired through contaminated soil and water.

Symptoms of Leptospirosis in Humans

The incubation period may be as short as four days or may extend for three weeks (in humans it is usually ten days). Fever, headache, chills, vomiting, weakness, muscle pain, and conjunctivitis (inflammation of the eyelids), or a combination of these symptoms are noted early in the disease. In the more severe infections, jaundice, kidney damage, blood disorders, and hemorrhages of the skin and mucus membranes occur.

Prevention

Avoiding exposure to the urine of wild animals, either in handling them or coming into contact with places where they have urinated, is the primary course of protection. Protective clothes, boots, and gloves are recommended. Good personal hygiene following any potential exposure is important.

Salmonellosis (Salmonella Enteriditis)

This disease is carried by reptiles, birds (pigeons, house sparrows, mallard ducks), and mammals (rodents, opossums, white-tailed deer). Salmonellosis is possibly the world's most common zoonotic disease, closely associated with poultry and swine industries. Humans comprise the next most important reservoir. Our immediate environment, such as landfills, sewage treatment facili-

ties, and meat processing plants, supply ready access to wildlife for contamination. Transmission occurs by ingestion of salmonellae shed in the feces and animal handling without using personal hygiene.

Symptoms of Salmonellosis in Humans

The disease is commonly manifested, after a short incubation period of six to seventy-two hours, as an enterocolitis (inflammation of the intestines) with sudden onset of headache, abdominal pain, diarrhea, nausea, and vomiting. The signs can vary from very severe to unapparent depending on the infecting dose, the patient's age, immune status, and stomach acidity. Dehydration, especially in the young, can be life-threatening. Fever is almost always high, and diarrhea can persist for days. Salmonellae can be shed in the feces for days or weeks. Antibiotics tend to increase the length of shedding.

Prevention

Good personal hygiene and proper environmental disinfections are important steps in the prevention of this disease.

Lyme Disease (Borrelia Burgdorferi)

The white-footed mouse and the white-tailed deer carry Lyme disease. Discovered and named a disease in the early 1980s, there are more than ten thousand cases reported each year in the U.S. Lyme disease has quickly gained the distinction of being the most common tick-borne disease in the U.S. The majority of cases occur in the northeastern seaboard states and the upper Midwest and California. Because white-tailed deer are increasing in numbers, it is probable that the geographic distribution of tick populations and, consequently, Lyme disease cases will increase in the U.S.

Lyme disease is a bacterium (borrelia) transmitted to humans by ticks that become infected by feeding on diseased animals, including birds, deer, rabbits, and rodents.

Symptoms of Lyme Disease in Humans

When a tick feeds on a human, it must be attached for at least twenty-four-to forty-eight-hours before infection can occur. The borrelia, residing in the tick's midgut, migrates to the salivary gland and enters the human's blood as the tick feeds. Upon inoculation of borrelia into the host's skin, an incubation period of three to thirty-two days follows before clinical disease develops.

In 90 percent of cases, initial symptoms include fever, muscle pain, lethargy, and the development of a characteristic skin lesion known as an "erythema migrans." Erythema migrans is a characteristic red annular "bull's-eye" lesion, which clears centrally as it expands. This lesion will spontaneously resolve, even without treatment. Untreated cases progress by being carried to many body sites: heart, muscles, bones, joints, liver, spleen, eyes, and brain. Signs usually involve the skin and musculo-skeletal and nervous systems. Headaches, muscle pain, and transient joint pains often occur. The majority of cases manifest as arthritis primarily of the large joints, especially the knee. Bouts of subtle nervous system disease such as memory loss, drowsiness, or behavioral abnormalities are characteristic. Case fatalities are rare, but patients experience a life-changing morbidity.

Prevention

Procedures to avoid contact with ticks are the primary recommendations. Wearing gloves, long pants, and long-sleeved shirts along with conducting frequent checks of clothing and body are recommended. When attached, remove ticks without crushing them and thoroughly scrub the wound.

In 1998 the Food and Drug Administration (FDA) reluctantly released a vaccine (LYMErix™) for Lyme disease, but researchers have many concerns about its use. So far only about 75 percent of those vaccinated have shown positive results.

Brucellosis (Undulant Fever)

Brucellosis is primarily centered in domestic agricultural species but may also be found in populations of bison, elk, and caribou in the western states. Brucella infections are usually transmitted by contamination of mucus membranes, especially by oral ingestion or through broken skin having contact with an animal's infected body fluids such as blood, urine, vaginal discharges, aborted fetuses, placentas, or feces. Aerosol transmission has also been documented.

Symptoms of Brucellosis in Humans

Incubation time is variable, ranging from one week to several months. In humans onset of the disease can be characterized by fever with accompanying headache, chills, profuse sweating, proration, insomnia, sexual impotence, or arthralgia.

Prevention

Protective clothing should be worn when handling these animals.

Zoonosis—Summary

- Zoonosis is a term that is used to identify any disease transported from vertebrate animals to humans.
- Vectors, such as ticks and mosquitos play an important role in the transmission of diseases.
- Any wild animal bite requires immediate medical attention.
- Wash any bite wound immediately and aggressively with soap and water.
- Traps and cages should be regularly cleaned and disinfected using a 1:10 mixture of Clorox™ and water.

- It is recommended that all trapped rodents be dusted with flea and tick powder prior to moving the animal.
- Always wear heavy gloves when handling caged animals.
- Always keep animal handling areas and human food areas separated.

Rabies

- Any warm-blooded mammal can carry rabies.
- Wildlife now accounts for more than 90% of all rabies cases.
- Early signs of rabies area characterized by apprehension, headaches, and a low-grade fever accompanied by vague sensory changes such as discomfort and irritation or pain in the area of the bite.
- Most cases of rabies are the result of bites from an infected animal.

Hantavirus

- The primary carrier of hantavirus is the deer mouse.
- Hantavirus is maintained in the deer mouse as a persistent infection characterized by long term shedding in the urine, feces, or saliva.
- Hantavirus undertakes a two to four week incubation period before disease ensues.
- Previously healthy individuals, when infected with hantavirus, suddenly develop symptoms of headache, fever, myalgia (pain in muscles) and dry cough.

Raccoon Roundworm (Baylisascaris)

- Baylisascaris is a disease process started by the ingestion of the raccoon roundworm eggs.
- Adult raccoon roundworms reside in the small intestine of raccoons. The adult worms produce and shed eggs in the feces.
- Adult raccoons are known to shed up to six million eggs per day.

- In the Midwest studies indicate that seventy to eighty five percent of the raccoons are infected.
- The most common manifestations are minor tissue damage and ultimate encapsulation of the parasite in non-critical sites such as skeletal muscle but can invade the central nervous system causing severe damage.

Bubonic Plague

- The primary carriers of the plague are prairie dogs and squirrels.
- Bubonic plague is primarily a flea-transmitted disease but can be acquired through direct contact with infected animals.
- After a two to six day incubation period of bubonic plague, an infection emanates from the site of the bite.
- With early detection, bubonic plague can be successfully treated with antibiotics.
- Case fatality rates reach fifty to sixty percent in untreated cases of bubonic plague.

Tularemia

- Tularemia is primarily carried by rabbits.
- Biting flies, ticks, and direct contact with the infected animal are equally responsible as sources of tularemia infection.
- The incubation time for tularemia is two to ten days.
- The most common symptom of tularemia is necrotic ulcer at the site of the bite, usually the hands. The organism then spreads to the regional lymph node.
- The case fatality rate in untreated cases is five to seven percent.

Rocky Mountain Spotted Fever

- Rocky Mountain spotted fever is the second most common tick borne disease in the United States.

- The primary vectors of Rocky Mountain spotted fever are rabbits, rodents, the Virginia opossum, the American dog tick, and the American wood tick.
- After exposure, there is a four to twenty four hour delay of the infection.
- Symptoms of the disease include lethargy and weakness, rapidly increasing fever, chills, headache, muscle pain, and pain in the joints. In eighty eight percent of the cases, spotting and blotchiness on the extremities and on the palms and soles is also a symptom.
- In the absence of specific treatment, a fifteen to twenty percent mortality rate occurs.
- Transmitted to humans by mosquitos and ticks.
- Incubation is usually 4 to 6 days after being bitten by an infected mosquito or tick.
- Early symptoms include headaches, high fever, pain and weakness in the muscles.

Encephalitis

- The most common hosts are birds, but the virus has also been found in rabbits and rodents.
- There are five common varieties of encephalitis in the United States.

Leptospirosis

- Leptospirosis is carried most frequently by striped skunks but can also be carried by raccoons, opossums, Norway rats, mice, and white tailed deer.
- Skunks can shed leptospires in their urine for up to 300 days.
- Transmission of leptospires occurs by direct contamination of mucus membranes or broken skin by urine.
- The incubation period for leptospires can range from four days to three weeks. The norm is 10 days.

- Symptoms include fever, headache, chills, vomiting, weakness, muscle pain, and conjunctivitis (inflammation of the eye lids.)

Salmonellosis

- Salmonellosis is possibly the world's most common zoonotic disease.
- Many birds and reptiles as well as rodents, opossums, and white tail deer carry this disease.
- This disease is closely associated with the poultry and swine industry.
- Transmission occurs by ingestion of salmonellae shed in the feces.
- Incubation period for the disease is six to seventy two hours.

Lyme Disease

- The white-footed mouse and the white-tailed deer commonly carry this disease.
- Ticks must be attached to humans for at least twenty four to forty eight hours before the infection can occur.
- Symptoms include fever, muscle pain, lethargy, and the development of a characteristic skin lesion known as an erythema migran (a "bulls-eye" lesion which clears centrally as it expands).
- Case fatalities are rare but patients experience life changing morbidity.

Brucellosis

- Carriers of brucellosis are domestic agricultural species, bison, elk and caribou.
- Brucellosis is usually transmitted by contamination of mucus membranes, especially by oral ingestion or through broken skin having contact with an animals infected body fluids.
- Incubation ranges from one week to several months.

- Symptoms of brucellosis include fever, headache, chills, profuse sweating, proration, insomnia, sexual impotence, or arthralgia.

Chapter Two

Animal First Aid

Some rescued animals will be incapacitated and will suffer from trauma. As with human injuries, first-aid in the field may be required to stabilize the patient. Without formal training, rescue volunteers should not attempt to perform any medical procedures but should be able to stabilize animals using basic first-aid. Finding a veterinarian close by who is willing to assist with medical emergencies will be very beneficial. Some veterinarians may be willing to assist during emergencies but reluctant to treat wild animals on a regular basis.

In many instances, it will be impossible to hold an animal down and give first-aid treatment without help. If you don't have the resources available to hold the animal securely and administer aid, wait until you get help. Trying to do both tasks at the same time can easily result in escape as well as injury to you and to the animal. Wild animals can be dangerous, especially if injured. This chapter is intended to provide the basic background information to allow you to successfully rescue, stabilize, and transport an injured animal to a medical facility.

Shock and Stress

Animals, like humans, are susceptible to shock. Shock is brought on by severe injuries of all types: hemorrhage, loss of body fluids other than blood, and stress. Pain, rough handling, and the delay in treatment can aggravate shock. The degree of shock is increased by abnormal changes in body temperature and by the animal's poor resistance to stress. Stress and shock can be fatal. To minimize shock in an animal, keep the handling of the animal to a minimum, cover the head and eyes with a towel, keep it in a quiet environment, and warm the animal as soon as possible. Unless the animal has life-threatening injuries, it is always best to give it time to settle down. Getting the animal warm and getting it hydrated can result in diminished stress and remarkable improvements to its overall condition.

Initial Exam

When approaching an injured animal, you must first make a general assessment of the animal's condition. You must determine if the animal is stable enough to undergo an exam, and you must attend to life-threatening injuries immediately. Observing movement and animal behavior can be very educational. For example, if the animal has been struck by a motor vehicle and you observe that the animal is dragging its hind legs, you have an indication that the animal probably cannot be rescued. Instead of giving a complete physical exam, you can check for pain responses and determine if more examination is necessary or if euthanasia is necessary. If, on the other hand, the animal appears to be moving its limbs but appears to only have skin abrasions, you will want to give it a closer exam. This will allow you to identify the type of wounds present and allow you to plan out the required treatments. Once you start your exam, start at the head and work your way back. Be consistent in your exam procedure, and you will be less likely to overlook potential injuries. As part of your exam, you should check for hydration (see fluid therapy) and check the animal's temperature if possible. Don't let obvious injuries capture your attention to the point where you stop looking and miss other not-so-obvious problems. Finish the entire exam before starting treatment.

Wounds with Severe Bleeding

The most common injury to animals is wounds. Wounds from vehicles, cats, dogs, guns, and other wild animals are common. Many of these wounds will require immediate first-aid. Severe injuries can cause dehydration and sudden temperature loss, leading to shock. Always keep this in mind when treating wounds.

If the animal is severely bleeding, you must take steps to stop the hemorrhaging immediately. This can be accomplished by applying direct pressure by hand over a dressing onto the wound site. It is important to use a dressing between your hand and the wound because the dressing will help control the bleeding by absorbing the blood and allowing the blood to clot. Do not remove this dressing after the bleeding is under control or the bleeding may start again. Wrap the dressing with vet wrap to keep the dressing in place. Be careful not to wrap too tightly or you may cut off circulation.

If the bleeding does not stop, you may be able to find the artery that is feeding the wound and apply a pressure-point bandage. A gauze pad rolled up, placed on the artery above the wound and wrapped tightly with vet wrap, may stop the bleeding. If applying a pressure bandage, do not remove the pressure compress from the wound.

In the case of a severe bleeding wound, do not attempt to clean the wound. Control the bleeding and transport. During transport, do whatever you can to immobilize the wound area and keep the animal warm and as calm as possible.

Wounds without Severe Bleeding

Wounds that are not bleeding severely should be cleaned before transport. Remove any foreign material from the wound and then clean in and around the wound by flushing the wound. Use a large syringe without the needle, filled with sterile water, Nolvasan®, or a saline solution. Always position the wound so that the flush solution flows away from the injury. Do not use iodine, alcohol, or hydrogen peroxide as your initial flush. It is best if the wound is moist when it arrives at the care facility. Apply a clean, dry pad over the wound and hold in place with vet wrap. If the wound is deep and will require stitching, a moist gauze pad can be placed against the wound to keep the tissue moist. Do not use topical ointments in the field because they can complicate future treatments.

Bites

The bite from another animal, whether wild or domestic, may result in an open or puncture wound. Dog and cat bites are common. Although a dog bite is likely to cause more extensive tissue damage, the cat bite can be more serious because a wider variety of bacteria are usually present in the mouth of a cat. Any animal bite carries a high risk of infection. In smaller animals (chipmunks, squirrels, and rabbits), infections can spread very quickly and cause death within a few hours. When a small animal suffering from a bite is rescued, it is important that antibiotics be administered as soon as possible. Clean puncture wounds by flushing as described above. Do not use topical ointments. Bite injuries should be transported to a care facility as soon as possible.

Gunshot Wounds

Gunshot wounds are treated very similar to puncture wounds. Bullets can cause severe bleeding that must be managed with pressure bandages and pressure-point bandages. If the wound is not bleeding, flush the wound extensively and cover for transport. Do not attempt to remove the bullet.

Fractures

If it is obvious that an animal is suffering from a broken bone, try to immobilize the broken limb using a splint and transport the animal. For a compound or open fracture, use a compress to stop any bleeding. Remove any debris and flush the wound. Secure the splint above and below the break. Do not attempt to set a bone or push a protruding bone end back into place. This will likely compound the injury. If the animal exhibits signs of a broken back, dragging hind legs, and an inability to move hips or legs, the animal will probably need to be put down. To check for a broken back, squeeze a toe on the animal's rear foot. If the creature does not reflex to the pinch, it probably has a broken spine.

Abdominal Injuries

If you are rescuing an animal with severe abdominal injuries, you should place a bandage over the wound and secure with vet wrap. Keep the vet wrap tight but not so tight that it causes constriction. Do not try to replace protruding intestines or abdominal organs. If there are protruding organs, moisten the bandage before wrapping to keep the area moist. Transport these animals as soon as possible to a care facility. As with any other injury, always cover the animal's head to reduce stress.

Eye Injuries

Eye injuries should not be treated in the field.

Burns

Animals with severe burns are difficult to treat. Burns cause extreme pain and dehydration and are highly susceptible to infection. Burn treatment is a long and painful process. Euthanasia is always a consideration when dealing with severe burns.

The first step in burn rescue is to examine the animal to determine the level and extent of the burns. There are three degrees of burns.

First-degree burns are minor burns only affecting the epidermis. First-degree burns are painful but not life-threatening. When treating these wounds, check the animal for hydration and then apply Silver Sulfadiazine Cream (SSD). Hydrate if necessary (see fluid therapy) and transport. Do not use home remedies (ice, butter, etc.).

Second-degree burns are burns that affect the epidermis tissue and the subcutaneous tissue. This level of burn is very painful and has even more potential for infection. When treating these wounds, check the animal for hydration and then apply Silver Sulfadiazine Cream (SSD). Hydrate if necessary (see fluid therapy) and transport.

Third-degree burns are the worst and are often difficult to treat in wild animals. These are full-thickness burns through several layers of skin and muscle tissue. The skin is blackened and many of the nerves are destroyed. If the animal is suffering from third-degree burns over large areas, euthanasia will be the best option. If the decision is made to transport these patients, cover the burned area with a clean, dry cover and transport. Do not hydrate or use SSD on third-degree burns because it will complicate future treatment.

Hyperthermia/Hypothermia

Raising or lowering the normal body temperature in nonhibernating animals can cause dangerous changes in the physiologic and metabolic processes. For the animal rescue professional, it is important to understand how important it is to maintain correct body temperature and the importance of quick mediation.

An animal's body temperature is maintained within a normal range called the set point. Hyperthermia occurs when the body temperature goes above the set point. Hypothermia occurs when the body temperature goes below the set point. Both hyperthermia and hypothermia can be fatal if not treated.

There are two types of hyperthermia.

Infections and inflammatory disorders can cause an increase in body temperature by generating heat via fever and increased metabolism. This temperature increase could indicate that the body's immune system is fighting an infection in an attempt to return to its set point.

This is a physiologic effect and unless at a critical level (>106°F), the temperature should not be forced down because this will interfere with the body's natural healing process.

Hyperthermia can also result from heat loading. Heat loading results when the animal is subjected to more heat than its body is able to dissipate. This condition can be caused by a hot environment, increased physical activity (trying to escape rescue), heat stress, heat exhaustion, or heatstroke. Heat load hyperthermia can and should be treated. It is sometimes difficult to determine which hyperthermia you have and when to treat. A basic rule of thumb is to keep any overheated animal on a surface that is cool and take no further action if the body temperature is less than 102°F. If the temperature is between 102°F and 105°F, put the animal on a cool surface in a cool, shaded environment and use a fan if one is available. If the body temperature is above 105°F, you should cool the body with alcohol wipes and a soft spray of cool water. If available, the animal should also receive a balanced electrolyte solution intravenously. To avoid shock, do not immerse the animal in cold water.

Hypothermia in wild animals is usually caused by exposure to unusually cold temperatures. This condition is often compounded if the animal is sick or injured. When an animal's body temperature reaches 82°F, it is no longer able to recover to its set point without intervention. When an animal's body temperature drops to 75°F, it will die within a few hours.

Nonhibernating animals can suffer hypothermia anytime their temperature drops below 99°F. The signs of hypothermia include altered mental state, depression, shivering, slow and shallow respiration, and cardiac arrhythmia (altered heart rhythm).

If any of these symptoms are observed, check the skin to see if it is cold. If the body seems cold, take the animal's temperature if possible.

To treat hypothermia, you must raise the body temperature as quickly as possible. Immerse the animal in a warm bath if available. If no bath is available, wrap the patient in warm blankets or towels. Use hot water bottles and heating pads if available, being careful not to inflict thermal burns. The patient should also be given a rapid intravenous infusion of a warm, balanced electrolyte solution as soon as possible.

Myiasis

Myiasis is defined as being infestation with fly maggots. Rescuing animals with maggot infestations is fairly common. Any open wound will attract flies, and flies are prolific egg layers.

The fly eggs will hatch within twenty-four hours and will become larva in as little as three days. Maggots burrow and feed on their host and grow quickly. These creatures, like most parasites, can have a devastating effect on the host. Fly larva like heat and moisture and flourish in open body cavities, and they tend to burrow into the subdermal.

When rescuing an animal with a wound with maggots, the wound must be flushed. Use a large syringe filled with sterile water or saline solution. Always position the wound so the flush solution flows away from the injury. Do not use iodine, alcohol, or hydrogen peroxide as your initial flush because these compounds can destroy tissue. Wash away as many maggots as possible. Because maggots migrate to subdermal areas, they will continue to appear in the wound. After the initial flush, the best way to rid the animal of maggots is to pick them out of the wound one at a time. After removing as many maggots as possible, cover the wound with a clean cover and transport.

There are medications that will kill maggots (Ivermectin, Capstar, Frontline and Revolution) but administration of these meds is best left up to the rehab staff or a veterinarian.

Fluid Therapy

Body fluids are vital for the survival of all animals. Without the proper amount of fluids, the body's internal transport mechanisms will cease to function and the animal will die. When an animal suffers from injury or sickness, it is likely to also suffer from shock and dehydration. Under normal conditions, animals will naturally eliminate fluids at the rate of about 5 percent of their body weight per day. These fluids are replaced daily from food and water ingested by the animal. A healthy animal will naturally maintain a balanced fluid level. Water makes up about 60 to 80 percent of body weight. It is distributed in the body as extracellular fluid (outside the body cells) and as intracellular fluid (within the body cells). It is also found in the blood plasma space.

If an animal becomes injured or sick, the rate of fluid elimination increases, and the animal becomes dehydrated and can quickly reach critical levels. In this case, fluids must be replaced.

Determining Hydration

There are several factors that must be considered prior to administering fluids to an injured animal. To avoid shock, highly stressed and critical patients should be warmed and stabilized prior to hydration therapy. To determine the extent of dehydration, pinch the skin in the middle of the animal's back into the shape of a tent (called tenting) and then release the skin. If the skin returns quickly (< 1 second) to its original position, the hydration level is high. The longer the skin takes to return to its original position, the more dehydrated the animal is. If the skin remains in the tent position, dehydration is at a critical level.

There are three steps required for fluid replacement.

1. Estimate the fluid loss. This estimate is made by observing clinical signs.

% Loss	Clinical Signs
<5%	No detectable abnormalities. Skin tent <1 sec.
6%	Very slight loss of elasticity (tent return ~1-2 sec.)
8%	Ten return 1-3 sec. Dry mucus membranes, dry gums.
10-12%	Tent return >3 sec. Lethargic, increased heart rate, cool extremities
12-15%	No tent teturn. Shock, very dry gums and eyes.

2. Calculate fluid replacement needs. This is done using the following formula.

> Animal's ideal body weight* in kg X % hydration = amount of fluids (liters) needed.
>
> *Ideal body weight = actual body weight + 10 percent of actual body weight

For example, a 10kg raccoon, 8 percent dehydrated would have an ideal body weight of

10 kg + 1kg (10% of 10kg) = 11kg

Using the formula

11kg x 8% = 0.88 liters of fluid needed

Replacement fluids should be administered in three sessions over a twenty-four-hour period. If the hydration percentage is 10 percent or greater, you should front load the fluid delivery. To front load, give 50 percent of needed fluid in first session, then two later sessions at 25 percent each. This first administration can be given in the field at the time of rescue.

After the initial day of fluid replacement, the animal will need maintenance fluids each day until its condition improves. Maintenance fluids are given three times a day. The amount of fluids needed is calculated using the following formula:

Animal's ideal body weight* in kg X 5% = amount of fluids (liters) needed

Using the same example as above:

11kg X 5% = 0.55 liters of fluid needed

Helpful conversions
1kg = 2.2 pounds = 1000g
1 pound = 0.454kg
1 liter = 1000 ml
1 oz = 30 ml

3. Determine the type of fluid to administer and the method of administration.

There are four common methods of fluid administration: oral (PO), subcutaneous (SQ), intravenous (IV), and intraosseous (IO). The method is usually dependant on the severity of dehydration.

When possible, PO should be chosen as the preferred method. PO is the least invasive method and is especially effective for cases where the estimated dehydration is less than 7 percent. Solutions given PO should contain sodium, glucose, bicarbonate, and amino acids and do not need to be sterile. Some common choices for PO administration include Pedialyte, Normosol, lactated ringers (LRS), and Gatoraid (diluted 50 percent). Fluids should be administered at body temperature. PO should not be used if the patient is vomiting, has dull senses, or has significant dehydration.

SQ is the injection of fluids with a needle or catheter under the skin. This is the easiest method and is recommended for rescue personnel. Fluids can be administered in large volumes (do not exceed 25 ml/kg) at multiple injection sites. SQ fluids are generally absorbed within eight hours. SQ fluids should contain isotonic concentrations of electrolytes. Some of the most common SQ fluids include lactated ringers, normal saline 0.9 percent, and Normosol. A 5 percent dextrose solution can be added to LRS in a ratio of 1:3 to provide extra calories.

IV administration is the administration of fluids through a catheter directly into a vein. Animals in shock or suffering from very severe dehydration should be treated using this method. IV catheters can be placed in peripheral veins or directly into the jugular vein. This is a sterile process and should be administered by a veterinarian.

IO is the most rapid method of introducing fluids. It utilizes a spinal needle and injects fluid directly into the bone marrow. Bone marrow provides a direct route to the circulatory system. This method is a sterile procedure, very invasive, highly susceptible to infection, and should only be used by a veterinarian.

Illnesses

Animals can carry a large number of diseases. Disease awareness and control is a very important part of animal rescue. Zoonotic diseases (diseases transported from vertebrate animals to humans) are always of concern to the rescue professional. Keeping track of any diseases detected in the area where you are rescuing and the symptoms of these diseases will give you a good heads-up. Most state health departments will have current information on any reported zoonotic illnesses.

There are also many animal diseases that are not zoonotic but can be easily spread through animal populations. Some of the most common and the most contagious diseases encountered are distemper, and parvo. Extreme caution must be used when these illnesses are encountered.

The symptoms of these and some other contagious diseases are confusion, walking in circles, falling over, runny nose, and caked eyes.

If you suspect the possibility of a disease in the animal you are rescuing, there are several precautions you must take. First, protect yourself. Wear gloves and approach slowly. Sick animals are totally unpredictable. If they present the above symptoms and are out of control, consider euthanasia. Remember, if you can approach a wild animal, something is wrong with the animal. If you don't see any injuries or have any idea why the animal is not attempting to escape, consider illness. Taking any rescue action that can cause injury to the rescue personnel or promote the spread of disease is not acceptable.

Highly infectious diseases are very easy to spread. Gloves, towels, catch poles, and carriers can easily become contaminated. If a sick animal is picked up, it must be isolated from all other animals. Any equipment that comes into contact with the animal must also be isolated. Be sure to advise the care facility about any suspected illness so they can isolate the creature. Contaminated equipment must be disinfected before further use. Soaking all contaminated equipment in a 10 percent Clorox solution can do this. Towels and other expendable materials should be disposed of.

Euthanasia

Euthanasia is the deliberate ending of the life of animals with incurable terminal illnesses or unbearable suffering. It is defined as an easy, quiet, and painless death.

Unfortunately, euthanasia is a necessary part of animal rescue. Animals that require rescue are, by definition, in some kind of trouble, usually the result of contact with humans. Traps, guns, cars, lawn mowers, fires, and domestic pets are examples of some human tools causing animals grief. Many animals can be fixed and returned to the wild, but unfortunately some cannot. The ones that cannot go back become euthanasia candidates. There is much debate about the euthanasia decision process.

On one side, there are those who firmly believe a wild animal can only have an acceptable quality of life if free to be in the wild. Keeping a wild animal in captivity is totally unacceptable. This group often considers the pain of treatment and the length of time to recovery. If pain is part of the treatment and the recovery time is too great, euthanasia is chosen.

There are also those who feel an animal can live a happy and content life in captivity. They believe life is always better than death when provided with adequate housing, food, and a low-stress environment. When used as educational animals, they also provide a benefit.

It is unlikely that people will ever agree on a standard protocol, but what is certain is that anyone who rehabs wild animals will at some point be faced with this decision. There are two rules with which almost everyone does agree. If an injured animal has a broken back or if there is a real threat of spreading diseases, the animal should be put down.

The most common and acceptable method of euthanasia is through the administration of chemical agents. Barbituric acid derivatives are common. Pentobarbital sodium is the most common and works by depressing the respiratory and vasomotor centers, causing heart activity to cease within minutes. Pentobarbital is a controlled substance and must be administered by a veterinarian.

Other methods that can be used include quickly breaking the neck of smaller animals, shooting the animal, or using a gas such as carbon dioxide or isoflurane.

First-Aid Kit

When rescuing animals, you should carry a basic first-aid kit for in-the-field treatments. The following items are recommended:

1. A variety of sizes of compresses. Four-inch compresses are about the right size for most animal injuries
2. An assortment of plain absorbent gauze and gauze roller bandages
3. Scissors
4. Sterile water
5. Saline solution
6. Nolvasan®
7. Lactated Ringers Solution (LRS)
8. Rectal thermometer
9. Assortment of syringes
10. Vet wrap
11. Several sizes of splint boards
12. Adhesive tape
13. An assortment of towels
14. Silver Sulfadiazine Cream (SSD)

Chapter Three

Rescue

This chapter will provide you with information on how to handle and rescue wild animals in a safe and efficient manner. Rescuing wild animals is a rewarding activity, but it is also dangerous. Becoming proficient and effective at rescue cannot be taught in a classroom. It requires many hours in the field gaining experience. Every animal rescue is unique, and every animal is unpredictable. There are no techniques that will work every time. There are, however, a few basics that can increase the probability of a safe and successful rescue.

The most important aspect of wild animal rescue is safety of the rescuer. This protocol can never be violated. It is far better to lose an animal than suffer a personal injury.

Besides causing pain, possible disease, and serious infections, most states require that wild animals that bite people be destroyed. Getting injured during a rescue will clearly make a bad situation worse.

The more you know about an animal species, the easier it will be to interact with the animal. Knowing how animals protect themselves, how they escape danger, what diseases they can transmit, and other behavioral tendencies will make the rescue easier and safer.

A majority of animal rescues involve animals that are either injured or sick. As people continue to expand into animal habitats, animal encounters will increase. Many animal injuries result from interactions with people and domestic pets. Cats are natural hunters, and dogs love to chase creatures that invade their space. Automobiles are extremely hard on wild animals. Lawn mowers, yard equipment, watering systems, and general household activities also take a toll. Animals that have been traumatized often need rescue. Some animals needing help, however, show no signs of injury. These animals may be sick. Both sick and injured animals are frightened, in pain, and feel totally vulnerable. They can be very dangerous and will do anything to protect themselves, including biting, scratching, and attacking.

On occasion, it won't be clear if the animal is in need of help or is just exhibiting a lack of fear of humans. This is especially true of raccoons and coyotes that thrive around people. One way to determine if the animal needs rescued is to slowly approach it. If you can get close enough to the animal that you can touch it with your foot, you can usually be assured that it is either sick or injured. Even animals that are unafraid of people will not let you get that close. An exception to this is abandoned baby animals. Babies in the nest can easily be approached. You must take care to ensure that an animal, especially a nest of babies, is in need of rescue. The last thing you want to do is rescue an animal that does not need help. Baby animals are much better off with their mother. The probability of an animal staying alive drops considerably when humans take over their care. When attempting a rescue, don't rush, especially if rescuing abandoned animals.

Always begin any rescue by standing back and watching. If it's a nest of babies, they may not be abandoned. Almost all baby animals needing rescue will demonstrate clear signs of distress. They will cry and wander out of their safe environment as if searching for Mom. If the babies are in the nest, are quiet, and seem content, you can assume they are not in distress. Nocturnal animals will usually stay with the nest until dark and then leave the nest to feed. Finding a nest of unattended baby animals at night is not uncommon.

General Animal Handling

Wait, watch, and evaluate.

When approaching a rescue scene, it is important to fully understand the situation you are entering. Sometimes the situation allows little time to wait because the animal may be trying to flee or it may be in a situation where additional injuries will occur if you don't act swiftly. In these situations, a quick response may be necessary. If time does permit, however, taking the time to watch, evaluate, and formulate a plan for capture can be very advantageous. When surveying the situation, there are several things to be considered.

Does the animal really need to be rescued? As already discussed, this is critical with babies. In the case of the cottontail rabbit, for example, the mother is seldom around her nest. She visits the nest a couple times a day, feeds her babies, and then leaves the nest. She spends the rest of her time hiding and watching. She does this to keep from attracting predators. Take the time necessary to convince yourself that the animals are truly in need before violating the nest.

Does the animal show any obvious signs of trauma? Is it going to require first-aid? If this is the case, the animal is not only scared and stressed, but it is also probably in pain. This combination makes the animal even more dangerous, and your caution level should increase.

Are there other animals or threats in the vicinity of the wounded animal? Animals that live in a family social structure often stay with injured family members. Mother animals seldom leave an injured baby, and your intrusion could provoke an attack. Domestic animals in the area can also raise the threat level if allowed to intimidate the injured animal. Carefully survey the area before approaching the injured animal.

Does the animal appear to be sick? There are many diseases that can incapacitate a wild animal. Some of these diseases can be transmitted to humans. An animal walking in circles, falling down, and appearing confused are all common signs of illness. A runny nose and infected eyes are also good indicators of illness. These signs, however, can also indicate head trauma. If you suspect an animal is sick, take extra precautions in handling the animal. Never approach a suspected sick animal without protective clothing. Keep clear of teeth and body fluids. Some diseases, like distemper, can spread very easily from animal to animal, and your equipment can be the vector. If you rescue an animal that you believe to be sick, isolate everything that comes into contact with the animal, including gloves, hoops, carriers, and vehicles. These items must be disinfected before they are put back in use.

Formulate a plan for rescue. A rescue can go very badly if it is rushed and not thought out. For example, if you capture an injured animal and you don't have a carrier ready or you capture it and need to stop massive bleeding but your medical supplies are still in your vehicle, you have a problem. Plan your approach, your equipment needs, your medical needs, your carrier requirements, and your transport plans *before you approach.* It is important to keep focused on the task at hand. If you divert your attention from the rescue, the animal gains the advantage.

Establish a Set Procedure

If you have a set procedure you use on every rescue, the chances of error diminish. Although every rescue may be a little different, if you are in the habit of doing things in a way that has proven successful, you will make fewer mistakes. The following rescue procedure is recommended.

After locating the animal:

- Wait, watch, and evaluate.
- Formulate a plan.
- Approach the animal.

Anytime you approach an injured or sick animal, you should move slowly and gently. Don't make any sudden movements, which will cause the animal to panic. If carrying equipment, don't wave the equipment around or allow it to spook the animal. If the animal begins to panic, stop and let it calm down, if possible. As you approach, if the animal becomes overly aggressive and you decide you need help, stop the rescue, back away, and get the help.

If the animal you are attempting to rescue becomes alert and aggressive, you may need to solicit help.

Once you approach the animal, you must decide which restraint tool to use. This depends on the extent of injury and how the animal reacts to your presence. Never assume an animal that appears to be totally incapacitated can be approached without danger.

Safety is the primary concern, but you must also always be concerned about reducing stress in the animal. If you are trying to capture a stronger, more aggressive, and dangerous animal (raccoon, badger, coyote, bobcat, etc.), you will usually want to use the hoop for capture. The hoop is also called a catch-pole.

The hoop is a valuable tool when attempting to capture large or dangerous animals. A hoop will cause stress to animals, but it does allow you to safely control them.

When using the hoop, there are two important rules to remember:

1. Once you decide to reach out and hoop the animal, don't hesitate.

When you reach for the animal with a hoop, it will move quickly to get away, it might attack the hoop, and it will often growl and bark. If you let these things distract you and hesitate with the "grab," you

A hoop will be one of your most useful tools.

will miss the animal. You must loop the animal around the neck and one arm (if possible), pull the loop tightly, and lock the cable in one smooth, continuous operation. This takes practice, and the target animal will never make it easy. If you miss the animal, each try thereafter will be more difficult.

2. Don't panic, even if the animal does.

Animals will usually become extremely agitated when you grab them with a hoop. Once you pull the cable tightly, it will do everything it can to get loose. It will scream, twist, flip, bite, scratch, and pull. You will think the animal is being tortured and killed. Keep focused and don't lose control. After you have the animal in the hoop, push it to the ground and pin it down.

If you reach for the animal with the hoop and catch it around the neck, you should continue the rescue attempt, keeping in mind that you are cutting off the animal's air supply and taking a chance at caus-

Putting the hoop around the neck can cause serious injury. Always try to hoop an animal around the neck and one leg.

ing serious injury. Hoops have been known to crush an animal's larynx. If you remove the hoop from its neck to attempt a better grab, you will probably lose the animal. If you hoop an animal around the neck, work quickly and make every attempt to get hand control of the animal as soon as possible so you can release the cable.

Once you have control of the animal and it is on the ground, immediately cover its head with a towel. With the towel over its eyes and the hoop still attached, hold the animal down as gently as possible until it calms. Do not release the hoop unless the animal is choking. Once the towel is in place and the action stops, it should calm quickly. Once it has calmed down, you can perform any emergency first-aid required. You can now place the animal into a carrier for transport. Use the hoop as a handle to move the animal. You can use the hoop to lift the front of the animal and with your other hand, you can scruff the animal on the back just in front of the tail. Place the animal into the carrier headfirst. This will reduce the chance that it will escape and will also keep the teeth farther away from you.

This is the safest way to contain an animal, but it also stresses the animal. Once you make your move to place the hoop on the animal, don't hesitate. This cannot be over-emphasized. The animal will panic. It will attempt to escape, it will scream, it will bite, and it will scratch. If you allow this panic reaction to cause you to back away, the chances for injury to you and the animal will greatly increase. Though the animal will act as though it is being killed, it is not. Get control of the animal and let it calm down.

For less-threatening animals (rabbits, skunks, rodents, foxes, etc.), you can often capture them with just a towel. This is the best method to capture skunks. If you hoop a skunk, it will spray. If you are gentle, approach slowly, and gently cover the skunk with a towel, you can sometimes avoid the spray. Always keep the animal in the towel when you pick it up to transport it. This keeps the animal's stress level down and gives you another layer of protection from teeth and claws.

A common error made by rescue personnel when carrying an animal is not keeping the animal away from the body. The captured animal wants to bite and doesn't really care if it bites your hand, your face, or your torso. Keep the animal as far away from the body as possible.

After the animal is in the carrier, you must warm the animal. Wild animals suffer shock the same as humans. Always warm it as soon as possible using hot water bottles, blankets, heating pads, or whatever is available that generates heat. When transporting, keep the carrier covered and keep the noise down. Stress and shock are killers to wild animals.

Rescue Equipment

Gloves

The most important piece of equipment you need when handling wild animals is gloves. Animals bite, scratch, spread disease, and carry parasites. You should never handle a wild animal without wearing gloves.

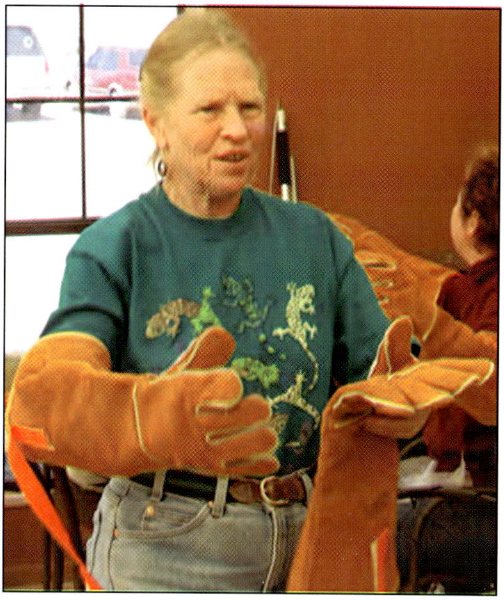

Heavy leather bite gloves are a necessity when handling wild animals. They come in a variety of sizes and styles.

There are three types of gloves you should have available for rescue purposes. Heavy-duty bite gloves are a necessity. Bite gloves are specially designed heavy leather-lined gloves that will repel most bites and scratches. Many brands of bite gloves are multi-layered with removable layers to give maximum protection and also allow for dexterity. They ride high on the arms for added protection from scratches. Some bite gloves also include a kevlar lining for more penetration protection. When selecting bite gloves for wild animal rescue, always search for gloves that are rated for large dogs and wild animals.

All bite gloves are manufactured with specific purposes in mind, and each kind has characteristics that either give them increased dexterity or increased protection. No one glove will meet all your needs, and all will have some limitations.

Medium-weight leather work gloves (or welding gloves) are also useful when handling cages and less-threatening animals. These gloves provide better dexterity, help prevent the spreading of diseases, and will repel bites from smaller animals. They are especially suited for smaller birds. Most mammals will be able to bite through a single-layer glove. These gloves can be purchased at any hardware store.

Surgical gloves are also a must when working with animals. For the seriously injured animals that don't pose a bite threat, these gloves will give maximum dexterity for administering first-aid and will provide adequate protection from blood and body fluids.

Gloves will provide you with protection, but no gloves on the market can guarantee you won't get bit. Raccoons, for instance, have very strong jaws and sharp teeth. If they get a good hold on your hand, they can bite through even the best bite glove. Coons can also crush fingers and break bones in the hand without actually penetrating the glove. Even with the gloves, use caution and keep your hand away from the animal's jaws.

The Net

This is an excellent tool for catching smaller injured animals and birds. It is safe and causes little stress. Many animals needing rescue have injured legs and wings. They can't go far, but they can go fast enough so that you will never catch them without an assist tool. A net with a long-reach pole will even the odds. When using a net, be sure after the capture to keep the net frame flat on the ground or you will probably be starting over. When the animal is in the net, approach the animal

The net is a valuable tool for small animals and birds. Be careful to keep the net flat on the ground after capture.

and place a towel over the net and the animal. Pick it up with your hands and place it in a carrier. Once in the carrier, remove the net but leave the towel with the animal.

Throw Nets

These tools are usually circular nets with weights attached to the outer diameter. They are used very similar to the way a cowboy uses a lariat. The net is thrown at the fleeing animal. These can be effective capture devices, but many hours of practice are usually required to become proficient. If you use a throw net and capture an animal, react quickly because the animal will escape quickly. Handle the animal using a towel as described above.

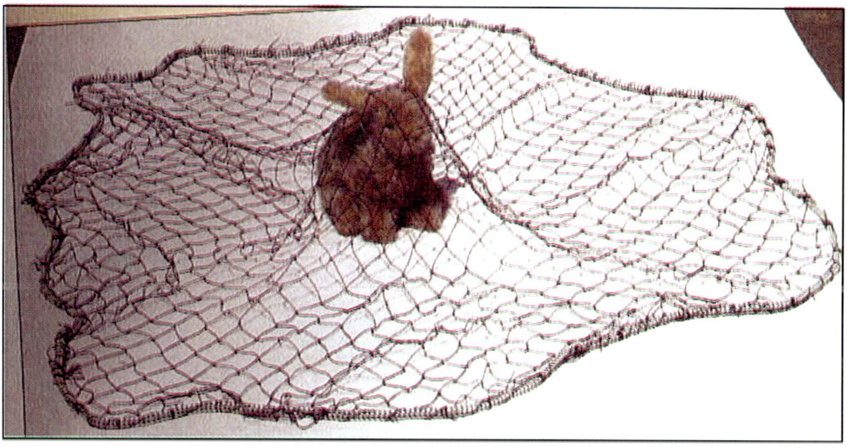

The throw net can be a useful tool, but practice is required.

The Net Gun

The net gun is a modified rifle that uses gas pressure from a blank rifle cartridge to propel a net toward the target animal. These tools are difficult to master and should only be used by trained and experienced individuals.

308-caliber Coda net gun, for hard-to-catch animals.

Trapping

Some animals, although seriously injured, are impossible to catch with capture tools. This is especially true of wounded animals that take refuge under permanent structures. These animals need to be trapped. Trapping injured animals can be difficult and tedious. It is suggested that when trapping is the only option, rescue personnel should consult with an experienced animal trapper.

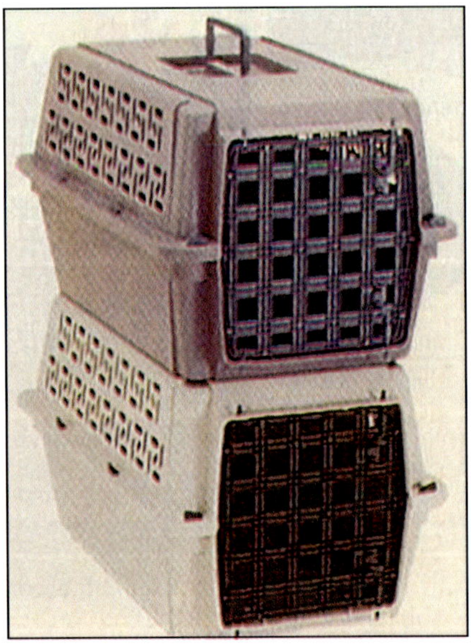

Pet carriers are the best transport containers.

Cages and Carriers

The best transport containers are the commercially available pet carriers. One large carrier is all you really need, but a variety

A suitcase-style carrier is ideal for small animals and babies.

of sizes are helpful. For raccoons and coyotes, you will need a large dog carrier. The larger the carrier, the easier it is to insert the animal. Some animals, like the raccoon, can grab the edges of any carrier and make insertion very difficult. A medium-sized carrier is better for skunks and smaller animals because it is easier to heat and prevent shock. It also restricts the area the animal has to thrash around, causing itself injury. For the same reasons, a small carrier is best for babies and small rodents. The suitcase-style carrier is great for babies because it is easy to heat and allows examination of the animals with a minimum of disruption and handling.

Towels

Towels are an absolute necessity when rescuing wild animals. Always carry a variety of towels with you on rescues. Towels have hundreds of uses in the field, from covering the animal to warming the animal, cleaning the animal, stopping bleeding, and adding extra protection to your protective equipment. Towels can be acquired from garage sales, thrift shops, and donations.

Dust Masks

It is recommended that you carry a supply of dust masks when rescuing. These masks should be used when working in enclosed areas and areas with animal fluids and waste present. These masks can block most infectious airborne diseases that are transmitted to humans.

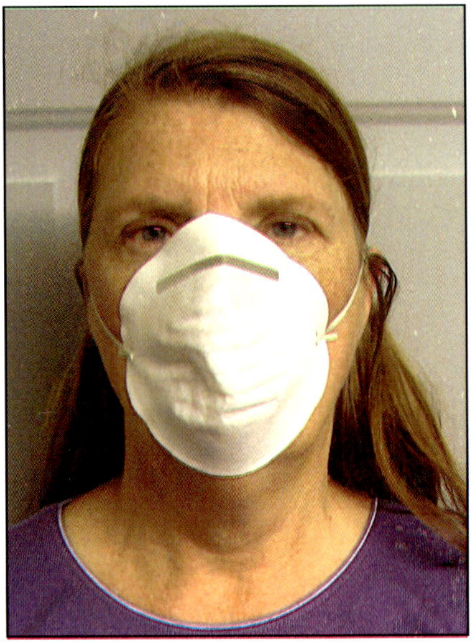

Always use a dust mask when working with sick animals and when working in confined areas and areas with animal fluids and waste.

Darts and Tranquilizers

Chemical restraints, although effective in some situations, are not practical for routine rescues. The tranquilizers in use today are designed primarily for clinical use. The problem with using them in the field is the fact that there is a significant time delay between injection and effect. An animal darted in the field usually has ample time to escape before the drug takes hold. It is then very difficult to find the animal. These drugs are also very tightly controlled and are not available to anyone except licensed veterinarians. If the situation arises where drugs are required, contact your local veterinarian and request help.

Chapter Four

Birds

Birds are another one of nature's favorite cat toys. They have found hundreds of ways to get themselves into trouble. A bird's main defense is its ability to fly away from danger. When it loses this ability, it becomes very vulnerable. Birds can be very fragile and require gentle handling when rescued.

When approaching a downed bird for rescue, it may fly away. If this happens, it is probably going to survive on its own. Birds that can fly still have a good defense and are almost impossible to catch. Birds that can't fly are usually easy to catch. Slowly approach the bird and cover it with either a net or a towel. Hold the bird down gently, and it should calm quickly. Pick the bird up, keeping the cover in place and making sure the wings are held against its body. If the wings are not held in place, the bird may injure itself trying to escape. If you are rescuing a raptor, be extremely careful of its talons. These long claws can seriously injure your hands.

Select a carrier for the bird that is just a little larger than its body. If the carrier is too large, the bird may injure itself trying to escape. A shoebox will work for most bird rescues. You do not need to punch holes in the box if you are transporting a short distance. The bottom of the transport box should be lined with torn newspaper strips (torn by hand and not a shredder). A thick layer of torn newspaper is needed to cushion the bird and prevent further injury. Do not attempt to give it any food or water unless directed to do so by a rehabilitator, and do not provide first-aid treatment to the bird while in the field. The most important thing to remember when rescuing birds is to keep them warm during transport. You can keep the bird warm by heating a towel in the microwave and placing the heated towel over the box. You can also place a hot water bottle in the carrier during transport, but be sure you cover it with a towel to keep the bird from sliding off.

Transport the bird as soon as possible. Stress is the number-one killer during rescue. Since flight is their primary means of escape, a bird that cannot fly is already in an elevated state of panic. When transporting, try to have a stress-free ride: no loud music, no loud talking, and turn down the air conditioner. Never transport birds in wire cages. Birds can get their wings through the bars and cause serious injury. For large birds such as crows, ravens, and raptors, a pet carrier is ideal. Put a rug or non-loopy towel on the bottom of the carrier to prevent the bird from slipping. Never transport more than one bird in each carrier unless you are transporting a nest of babies.

Birds do not carry rabies but can spread some diseases. Always wear gloves when handling a bird.

A Cat-Caught Bird

All cat-caught birds should be transported to a rehabber as soon as possible. The sooner it receives treatment, the better its chances for survival.

A Bird Hit My Window

Many of the birds that fly into a window and fall to the ground aren't seriously injured but are only stunned. If there is an obvious injury—bleeding from the beak, off-balance, broken wing, etc.—transport it immediately. If there are no obvious injuries, the bird may only be stunned from the collision. When rescuing one of these birds, place it in a small enclosed container and put the container in a place that is dark, quiet, and warm. After about two hours, take the container outside, open it, and allow the bird to escape. More often than not, the bird will fly away. If it can't fly away, transport it to a rehabilitation facility.

A Bird Fell Out of Its Nest

When a baby bird is found on the ground, it probably fell or jumped out of its nest. Before you make a decision on what you should do with the bird, you must determine what you are working with. Birds in different stages of their development have different needs.

A fledgling is a bird that is fully feathered but has very short tail feathers and is ready to leave the nest (has fledged from the nest). A fledgling will leave the nest on its own as a natural part of its development. It will spend a few days on the ground or in low bushes before it can fly and feed on its own. Usually you will find more than one in the same area. During this time, the parent birds are still feeding the fledglings and showing them where to feed. The parent birds cannot pick up the fledglings or move them to safe locations, so they are vulnerable. Most rescued fledglings are erroneously rescued. Try to keep pets and children away from the birds and watch from a distance to be sure the parents are feeding them. If no adult bird comes in two hours, rescue is appropriate. When watching, keep your distance or the adults may not approach the babies.

A hatchling is a newly hatched bird and will be featherless and have its eyes closed. These birds cannot regulate their body temperature and need immediate attention. When you rescue a hatchling, try to locate the nest from where it fell. If you can locate it, put the bird back into the nest. Folklore tells us that a bird that has been handled by humans cannot be returned to its mother. This is not true. Birds have a very poor sense of smell. Many can't smell at all. Babies that have had human contact are readily accepted back into the nest. If you can't find a nest or it's too high to reach, warm and transport the bird.

Nestlings are birds that have either feather traces or downer feathers. These birds also need warmth and care from the parents and should also be returned to the nest.

If you are rescuing a baby raptor, watch for the parents, who may be nearby. They may try to attack you if they think you are harming Junior. Anytime you return a bird to the nest, you should stay around and watch for one hour to be sure the baby stays in the nest and the mother returns. If she doesn't return to the nest in an hour, you should rescue all the babies and transport them to a rehab facility.

The Nest Fell Out of the Tree

Occasionally you will be called to a site where the entire nest of babies has come crashing down. In this situation, you should try to locate where the nest had been. If you find it, remove the baby birds to a warm and secure carrier and put the nest back. You can hold it in place with string or wire. When the nest is secured, place the babies back into the nest and watch for at least an hour. If the nest is damaged, you can repair it with string or wire and put it back into place. Be sure not to leave sharp points or loops in the wire, which might harm the birds. If the nest is severely damaged or destroyed, you can construct a new nest. Use a plastic flowerpot or similar container and line it with straw or the debris from the old nest. Don't use grass or paper because these materials will quickly mold. Be sure your nest has large drainage holes in the bottom. If you're not sure where the nest came down from, select a spot in the same general area and attach the nest with string or wire. Be sure the site you selected is not in direct sunlight and is sheltered from the rain. Watch from a distance to be sure the mother returns.

A Bird in the Chimney

It is common for birds to get themselves trapped in fireplace chimneys. The easiest way to get them out is to open the fireplace damper and block off the firebox to the room. Turn out the lights, music, and TV, and keep people and pets out of the room. The bird will usually find its way into the firebox. Once in the firebox, close the damper and capture the bird in a towel or net. Examine the bird for soot and creosote contamination. If the bird is clean, let it go outside. If it is heavily covered with soot, you should transport it to a rehabber for a bath. Fireplace soot is highly corrosive. Another method to get the bird out is to drop a knotted rope down the fireplace from the top. This will give the bird something to grasp and allow it to go out the top.

A Bird in the House

When trying to rescue a bird that is loose in the house, contain the bird to one room that has an outside exit and windows. Turn out the lights and open all the doors and windows. Using your net or a sheet of cardboard, try to herd the bird toward an exit. As soon as the bird sees an escape route, it will usually fly out.

A Bird Is Caught in Sticky Traps

If the bird is caught in a sticky rodent trap, you should remove it from the trap very carefully. These traps can easily tear the skin from a bird. Wrap the bird in a towel and transport. If you have any canola oil, you can try to remove the glue prior to transport.

A Bird Fell into a Puddle Of Oil

If you rescue a bird that has been caught in oil, keep it warm and transport it. Do not try to clean the bird in the field.

Birds Having Suspected Diseases

When you rescue a bird that appears to be sick, there are several common bird diseases that are easy to spot. Although you cannot treat the birds in the field, you can warn the homeowner of what you suspect and possibly avoid future occurrences.

Salmonella

Salmonella is a highly contagious disease. Birds with salmonella are usually found in water during the winter months, when the weather is cold and there is snow on the ground. At this time of year, people don't clean their feeders as often and they become contaminated. Salmonella is usually found in finches and evening grosbeaks (evening grosbeaks bite like parrots, so be careful). The symptoms of this disease are fluffed feathers, diarrhea on the feathers, and seed (like bubbling yeast) coming from the beak.

If you suspect this disease, notify the homeowner and suggest they clean and disinfect or remove the feeder and baths. Wash your hands and equipment after having contact with these birds.

Trichomonasias

This disease is usually found in pigeons and doves. A bird with Trichomonasias will be lethargic, weak, and will have cottage cheese-looking saliva coming out its beak during the advanced stage. This disease is also highly contagious. If you suspect this disease, notify the homeowner and suggest they clean and disinfect or remove the feeder and baths. Wash your hands and equipment after having contact with these birds.

Maggots

If you pick up a bird with maggots, warm it and transport it to a rehab facility.

West Nile Virus

This is a relatively new disease that is spreading rapidly. It usually affects crows and ravens. The symptoms of this disease are seizures, head tilt, stumbling, vomiting blood, and emaciation. If you pick up a crow or raven with these symptoms, you should contact the State Health Department to determine if they want to run tests on the bird.

Raptors

When rescuing raptors, you must use extreme care. The feet or talons on a raptor can cause serious injury. The beak looks threatening but is not the main threat. Always wear heavy bite gloves when handling raptors.

Raptors are also highly protected birds by both State and Federal law. It is illegal to harm or possess raptors. Any rescued raptor must be transported immediately to a licensed rehab facility. Although very large and robust-looking birds, they are actually fragile birds and should be handled gently.

When rescuing a raptor, approach the bird slowly, from the rear, if possible. If the bird is alert and can follow your movements, anticipate that it will struggle when first covered. When close enough, carefully place a towel or blanket over the bird, being sure to cover it completely. It may take several attempts. Remember where the feet were so you can avoid them. Carefully gather the cover together. The best approach is to slide your hands down from the back, pinning the wings to the body as you grasp a leg in each hand, just above the ankle. Lift the bird up against your chest, with the feet pointing away, and then wrap it firmly. Be careful because it will feel like you have a blanket full of talons. Remember, a raptor's feet and talons are its main defense.

Place the bird on its stomach in a large sturdy cardboard box or carrier that is about twice the size of the bird (if the box is too small, it can cause additional injury). Fill the box or carrier half full with torn strips of newspaper, which will act as insulation and padding and will also help support the wings and legs. If possible, use newspaper rather than a towel because large raptors can get tangled in a towel. The box should be well ventilated. Darkness will keep the bird quiet.

Baby owl.

Chapter Five

Snakes

Rescuing snakes can be a serious business. Unlike other animals, a venomous snakebite is always very serious and, if untreated, can be fatal. Before attempting a snake rescue, you should know what kind of snake you are approaching, know if it is venomous, and know how it is likely to behave. Without adequate knowledge and training, you should not attempt to rescue snakes.

Snakes that are found in our communities are generally not sick or injured but are in search of shelter, food, or water. During times of drought, snakes have to travel farther looking for food and water and as a result, more snakes are seen in backyards, garages, sheds, and even homes. The primary food source for most snakes is rodents (i.e., mice, rats, squirrels, and rabbits), which can carry zoonotic diseases. These diseases, however, only affect warm-blooded animals. Because snakes are cold-blooded, they cannot carry, transmit, or become ill from these diseases. Snakes are Mother Nature's perfect "mouser" because they hunt day and night and above and below ground. No other creature can control rodents like the snake.

Snakes are on the low end of the food chain, which means many predators eat snakes (i.e., roadrunners, birds of prey, coyotes, foxes, and bobcats). Snakes see humans as big, bad predators coming to hurt them, and they react accordingly. To protect themselves, they will coil, hiss, rattle, strike, and bite. These actions will often frighten people away. The biggest challenge faced in many snake rescues will be the freaked-out homeowners prior to the rescue.

Since snakes are cold-blooded, many slither onto highways or roadways to seek heat from the asphalt, especially during cool summer nights. Most snakes are killed this way. People who witness a snake being run over by a vehicle think this will not hurt the snake, but most will die a slow death. Be very gentle with these injured snakes, and transport them as soon as possible to a reptile rehabilitation facility.

There are over eight thousand snakebites in the United States each year. Most bites come from just a few species of snake. You should be aware of each of these species.

Snakes in the pit viper family (Crotalidea) account for about 99 percent of all venomous bites. Snakes in this family include the rattlesnakes, copperheads, and cottonmouths, or water moccasins. These snakes get the name "pit viper" because they have heat-sensitive pits just above their nostrils that allow them to locate their prey. Eastern and Western diamondbacks account for the majority of the snakebite deaths because their venom is one of the most toxic. Cottonmouths, or water moccasins, are aggressive snakes that live both in and out of water. They have medium-potency venom and primarily live in the southeast United States.

The coral snake, found throughout the southern United States, accounts for a minority of snakebites. These snakes are nonaggressive and, unless provoked, seldom bite. They have very small fangs, but unlike the pit vipers, they are not retractable. When they do bite, they transfer their venom by chewing rather than injecting. Their venom is medium toxic.

Pit Vipers—Rattlesnakes

The most effective ways to identify the rattlesnake are body coloring, markings, and the shape of the head and tail. Mature rattlesnakes have

Western diamondback rattlesnake. Most snake bite deaths are caused by the diamondbacks.

rattles and a flat triangular-shaped head. Newborn rattlesnakes do not have rattles, but they have a pre-button on their tail. A pre-button resembles a kernel of corn. Non-venomous snakes have a pointed tail, and the head is much more rounded. The diamondback also has a series of black-and-white stripes just in front of the tail. These stripes are very distinctive and are on both the Eastern and the Western Diamondbacks. Because of these distinctive stripes, some people refer to these snakes as coontail snakes.

Other common rattlesnakes:

- Banded rock
- Black-tailed
- Canebrake
- Massasauga (eastern and western)
- Mojave
- Mottled rock
- Pacific (northern and southern)

- Pigmy (southeastern and western)
- Prairie
- Red diamond ridge-nosed
- Sidewinder
- Speckled
- Tiger
- Timber
- Twin-spotted

Pit Vipers—Copperheads and Water Moccasins

The most effective ways to identify the copperhead are body coloring and markings. The copperhead has very distinctive copper-colored markings on its body. It also has a solid copper-colored head. The head is flat like the rattlesnake. This snake will be seen basking during the

Copperhead snake. These snakes are seldom aggressive.

day when weather is cool but not cold. During summer heat, it will stay hidden during the day and become active at night.

Common copperheads:

- Broad-banded
- Northern
- Osage
- Southern
- Trans-pecos

Cottonmouths are large, aquatic, venomous snakes. They are generally dark olive, brown, or black. A lighter/darker cross-banding pattern may be seen, especially on the sides. They are heavy bodied with keeled scales. Adult size is considered to be in the range of 30 to 48 inches. The young are more distinctly patterned. They also have a white-lined mouth that is very distinctive.

Cottonmouth or water moccasin. These snakes are large, very aggressive, and dangerous snakes.

Common Cottonmouths:

- Eastern
- Western
- Florida

Coral Snake

Due to the small size of coral snakes, along with having much smaller fangs than pit vipers, bites are frequently ineffective and have some difficulty penetrating shoes or even thick clothing. This, along with the fact that coral snakes are not aggressive creatures and reclusive, makes bites quite rare.

Common coral snakes:

- Arizona
- Eastern
- Western
- Texas

Coral snakes are most notable for their red, yellow/white, and black-colored banding.

Rescuing Snakes

To successfully rescue snakes, you should have adequate training and understand both the snakes and the danger of being bitten. Even a very slight bite from a rattlesnake can be very serious and can only be treated with antivenin (Cro Fab), which is administered at a hospital. An envenomation usually results in severe pain, loss of body tissue, or sometimes even death.

Snake identification is very important when attempting to rescue. There are many snake species in the United States, both venomous and non-venomous. The most common venomous snakes are described above. It is important to know what species are native to the area before attempting snake rescues. If you are not absolutely positive what the snake is, do not pick it up or handle it with your hands. Always wear thick leather boots and leggings/chaps and use hooks or tongs to handle/manipulate the snake. When relocating snakes, carefully place the snake in a lockable container for transport.

Don't rush the rescue. Carefully survey the area for other snakes or hazards. After you have located the target snake, try to identify it before you attempt the rescue. This is much easier to do if you are familiar with snakes, their coloring, and their habits prior to going into the field. For example, some snakes, like the bull snake, will mimic a rattlesnake by coiling, hissing, and striking out. Once you have identified the species, try to always keep it in sight and approach slowly. When approached, rattlesnakes will often give out a warning by coiling and rattling. Some snakes are very aggressive and will attack. Remember, a rattlesnake can strike to a distance of about half its body length. If the snake moves out of your sight, you will need to try and locate it again. Snakes can quickly disappear under rocks, brush, woodpiles, buildings, or in a hole. To relocate them, you may need to move or look under debris. <u>Do not</u> search or move debris with your hands. Use your hook or some other reaching device, and do not lose sight of your feet or hands.

Once you locate the snake, use your hook or tongs to move the snake into an open area where you can work without obstacles.

The hook is a pole with a heavy wire hook protruding from its tip. It does not have any moving parts and is used to simply lift the snake off the ground while allowing the user to maintain a safe distance. It is the best tool to use in handling snakes because it will not injure them.

Snakes move by contracting body muscles against the ground. If a snake is lifted off the ground at a single point along its body, it cannot move forward. This is how the hook works. It lifts the snake off the ground, immobilizing its movement and allowing it to be moved.

The tongs are a device with a mechanical set of closable fingers attached to a pole. The fingers are opened and closed with a trigger on the handle. They are used to actually grab the snake with the finger and then lift it off the ground. Tongs give positive control of the snake but can cause serious injury to the animal if griped too tightly. Snakes have

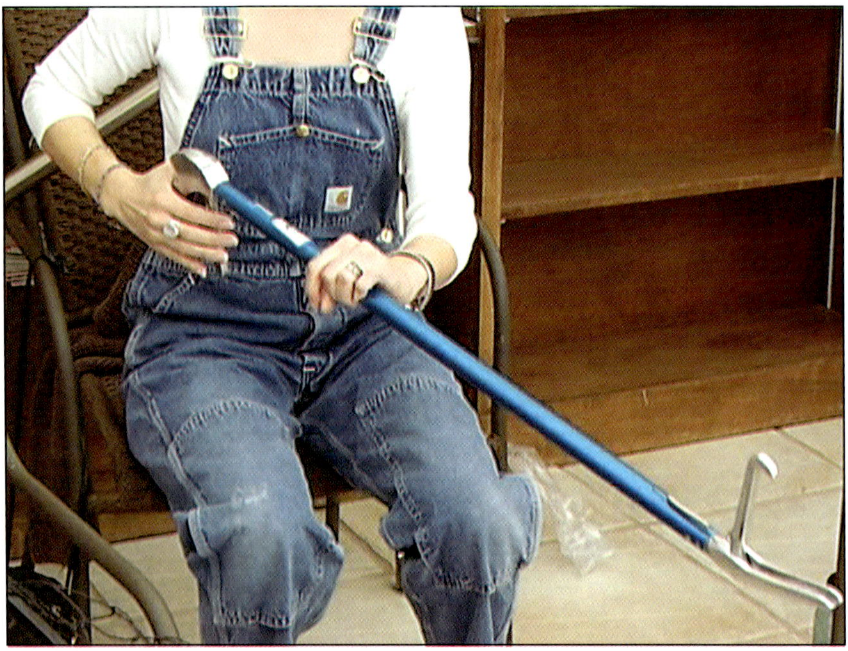

Snake tongs are easy to use and give more control over the animal but may cause injury to the snake.

Snake hooks are THE choice of experienced snake handlers. They are easy to use and do not injure the snake.

a very weak bone structure directly behind the head for about one-fourth the body length. This area should always be avoided when gripping a snake with tongs. When tongs are used on a snake, it becomes stressed when it feels the tongs grabbing, pinching, and sometimes crushing, and is more apt to bite the tongs or even itself. Picking the snake up with the hook is less stressful to the snake and easier for the experienced snake handler to have a successful, safe capture.

Once lifted off the ground, the snake loses its mobility and often becomes confused, making it much easier to handle. Before you lift the snake, be sure your carrier is ready to accept it. Place the snake into a secure container with a locking lid. The carrier should be at ambient temperature and dry to keep from stressing the animal. When closing the lid on the carrier, use your hook or tongs so your fingers do not become targets for a bite.

In Case of Snake Bite

If bitten by a snake, there are things you can do and there are things you should not do.

If you get bit, **DO**:

- Remain as calm as possible.
- Proceed to the nearest medical facility ASAP.
- Avoid physical activity and exertion.
- Immobilize the bitten area and try to keep below level of heart.
- Remove tight-fitting garments and constricting jewelry, like rings.
- Try to identify the snake accurately.

If you get bit, **DO NOT**:

- Make any incisions.
- Apply suction by mouth.
- Apply pressure or squeezing at the bite site.
- Use tourniquets or other constricting devices.
- Use ice or ice-water immersion.
- Inject anti-venom.
- Use stun gun or electric shock.

Chapter Six

Animal Behavior

Over the years, we have realized the more we know about our target animal, the easier the rescue. Animal information has not only helped us, but it has also prevented many unnecessary rescues. For example, we now know that when a baby raccoon gets separated from Mom during a night training session, the baby will stay where it is until Mom comes back, which may not be until the next night. The baby will go to sleep wherever it happens to be. People often find these little creatures sleeping in their backyard or on their porch and will call for help. It is obvious that the baby is out of place and at first glance, it is assumed that it needs to be rescued. Knowing, however, that Mom will probably be back after dark allows us to just observe the animal for another twenty-four hours without disturbing it. Mom always seems to show up.

The information that follows is a brief description of the normal behavior of some of the animals we rescue on a regular basis. Not all animals are going to react as anticipated, but knowing some basic behavior traits can sometimes give you the upper hand during rescues. We believe the more you know about the animal, the more successful you will be.

Raccoons

Raccoons are probably the most common backyard pests in the United States, and they frequently get into situations where they require rescue. They are found in every state and most of Canada and Central and South America. They are adaptable animals that thrive around humans. Raccoons are natural-water as well as land hunters and feed on almost anything available. Raccoons prefer mature woodlands and thrive along streams, but recent studies indicate that raccoons can survive quite well in more arid climates.

In most neighborhoods throughout this country, the animal's main diet consists of dog food, bird food, lawn bugs, garbage, and pond fish. The raccoon eats at night and is known for washing its food prior to eating. There are several theories about this behavior, but published studies indicate that raccoons are natural water hunters and when raccoons are feeding around water, they use their paws to locate and identify food. The touch sensitivity is much greater when the feet are wet. We have observed raccoons tell the difference between green grapes and red grapes by feeling them in water. When feeding where water is present, they submerge their food to feel it before ingestion. If no water is present, they just eat it.

Raccoons do not hibernate like their cousin, the bear. They do, however, go into what is called semi-hibernation during the winter months. A raccoon's metabolic rate drops, which allows it to go for several days at a time without food and water. To prepare for this, the raccoon builds a fat reserve during the fall months by

going on a feeding frenzy, focusing on the many fruits that ripen during the fall season. This can be important if the raccoon is out feeding during harsh winter weather. If the action does not follow the expected norm, illness could be a possibility.

Raccoons reproduce once a year, in the spring. The gestation period is sixty-three days, and the average litter size is four to six. Before people invaded their territory, raccoons had their litters in hollow logs, dead trees, or burrows under rocks and logs. Now that people have invaded their territory, raccoons prefer to have their babies under buildings, in attics, in fireplaces, or any other place they can make noise at night to keep the dogs barking. Raccoons are nocturnal animals, so they are out feeding at night. When they are nesting, they return to the nest in the early morning and spend the daylight hours feeding and caring for the babies. This results in a situation where the mother is awake at night and the babies are awake in the daytime. At eight to ten weeks, Mom will start the training process and let the babies tag along at night. This can be significant to the rescuer because the little guys are used to being awake during the daytime hours, which is contrary to what we know to be nocturnal behavior.

Predators of raccoons include mountain lions, domestic dogs, coyotes, and humans. They are also frequent targets for fur trappers and hunters.

Raccoons can be very vicious and aggressive when cornered. They can move quickly when defending themselves and can cause serious injury. The main threats posed by raccoons come from their teeth and their feet, especially the hind feet. When attacked by another animal, raccoons will often lay on their backs, making themselves look vulnerable. When the animal attacks the apparent submissive raccoon, it will pull the aggressor in with its front feet and tear at the animal's underside with its hind feet. It usually only takes a couple of blows and the attacker is killed. Because of this behavior trait, it is not uncommon to find wounds on the raccoon's underside. Around water, the raccoon will tempt its attacker into the water by appearing to be in distress. When its attacker enters the water, the raccoon will grab it and force it under water until it drowns. There are few animals that can take on a raccoon and win. Unlike many other animals, raccoons don't show a lot of fear of humans. It is not uncommon for homeowners to find themselves in a stare-down with a raccoon over a bowl of pet food. Some pet owners will strike out at the animals with brooms, sticks, pellet guns, and bullets. These are injuries that should be looked for. Raccoons will also seek the warmth and security of a household chimney to nest and give birth. The mothers and babies must be removed from the chimney because the fireplace contains chemicals that are caustic to the newborn animals and can cause skin burns and blindness.

Under normal circumstances, the raccoon is not an aggressive animal and will almost always give a warning prior to attacking. If the threat goes away, the raccoon will usually just go on about its way. When threatened, the raccoon will usually lower its head, hump its back, and let out a threatening growl. If the threat continues, the raccoon will leap forward with a loud barking sound. If this doesn't work, the raccoon will attack and bite.

When a raccoon is threatened, it will lower its head, hump its back, and let out a threatening growl.

Raccoons have sharp teeth like a dog's. Bites are usually deep and can be very painful. When it bites, it will often refuse to release its grip. Raccoons have very strong jaws and can cause serious damage. If they get hold of a hand or finger, they can easily crush bones. They are also very tenacious animals and will never give up their fight to get free. They will also strike out with their front feet, which can cause a painful injury.

When held down, a raccoon can generate a variety of blood curdling sounds, from barking to growling to loud screams. Even in extreme distress, however, the raccoon will continue to take advantage of opportunities to bite and scratch.

The best way to confine a raccoon while trying to rescue it is to cover its head and eyes with a heavy towel, scruffing it behind the neck while pushing the head down to the ground with one hand and using the other hand to push the body down. To pick the animal up, keep hold of the back of the neck and then scruff the animal on its back directly in front of the tail. If you can't get a good grip on the back, use the tail. To be successful, you must be firm and aggressive. The longer you hold the animal, the greater the chance it is going to get free or bite you. If you are rescuing a raccoon that needs emergency medical attention, you will need two people.

Getting the animal into a carrier can also be a difficult task because the animal will grab on to everything within reach to prevent its confinement. Again, to be successful, you must be firm and aggressive. If you let the raccoon convince you that it is being hurt and you back off, you will get hurt.

Raccoons at a Glance

Natural Foods	Fruit, small rodents, crayfish, insects, acorns, worms, eggs, roots.
Territorial	No. The raccoon is willing to share its home range.
Range	Male, several miles. Female, about a mile.
Hibernation	No. Feeds year round.
Reproduction	Once a year, April - May.
Gestation Period	63 days.
Litter Size	4 to 6 kits.
Leave the Nest	20 to 25 weeks.
Predators	Mountain lions, bobcats, domestic pets, and humans.

Skunks

In the animal kingdom, the skunk is probably the most notorious animal. Everyone recognizes the skunk and its powerful defense mechanism.

In reality, skunks are very slow-moving animals and are gentle, affectionate, and docile. They seldom get excited, hardly ever fight with other animals, share their den sites, and are quite intent on minding their own business. Their high-contrast black and white colors seem to advertise the fact that they are there and are not afraid of anything. When threatened, they discharge a scent that will stop almost any attacker. Adult skunks carry about a tablespoon of this noxious spray and can control the discharge rate, allowing them to spray up to five times in a very short period.

Some people think the skunk spray comes from the animal's urine. This is not true. It actually is emitted from the animal's anal glands

through tube-like ducts. The skunk can spray up to fifteen feet with very high precision. The spray itself is a sulfur-alcohol solution known as butylmercaptan. This spray is so potent that it can be detected for over one-half mile. Baby skunks cannot control their discharge and therefore discharge their entire supply of THE chemical in one shot. Skunks almost always give warning before they spray. They stomp their front feet, spin in circles, and raise their tail. If the threat does not disappear, the skunk will turn its tail toward the victim and spray. After the skunk releases its spray and the threat goes away, it will continue on as if nothing happened. Skunks spray only when threatened, and the females spray the males as a warning to stay away. In areas with no threat, they can live with their human hosts for years totally undetected.

Skunks warn before they spray. If you get this view, it's usually too late.

Skunks have very strong forefeet and long nails, making them excellent diggers. Much of the skunks diet is found by digging. Skunks will spend considerable time digging up earthworms in nicely planted lawns. Like other members of the mustelid family, skunks have small feet with five toes on each foot. There are three genera of skunks in the United States: the hog-nosed (conepatus), the spotted skunk (spilogale), and the most common, the striped skunk, which has two species,

Front- and hind-foot tracks of a striped skunk.

the common skunk (mephitis) and the hooded skunk (mecroura). The tracks are often confused with cat tracks. Cat tracks are the same size and shape but only have four toes and don't show claw marks. The skunk track almost always displays its long claws in its track.

When first born, baby skunks cannot spray. At five to six weeks, they begin to produce a musky odor and will be able to spray shortly thereafter. Baby skunks, unlike adults, cannot regulate the amount of spray they emit. This means the little ones can be just as potent as the big ones. Skunks have a territorial range of only thirty to forty acres. In the wild, they live in shallow burrows, in caves, or under logs. Since urbanization has occurred, the skunk has found suitable homes in and under buildings, under decks, in garages, and in woodpiles. They like dry, dark, and defensible areas, which are easily found in or around almost any home. Skunks are nocturnal and begin feeding at sunset.

Mating takes place from February through March. The gestation period varies by species, and the female gives birth to between four and eight young. The babies are born with their eyes closed and are completely helpless. They nurse for about two months. The mother skunk is fiercely protective of her young. Like the raccoon, adults feed at night and care for their young during the day. When the babies start emerging from the nest, they too can get separated. Baby skunks will stay put and wait for Mom to come get them. At about eight to ten weeks of age, babies will also sneak out of the nest during the daytime, while Mom is sleeping. There is a myth that if a skunk is out during the day, it has rabies. This is not true. It is not at all uncommon to see a juvenal skunk out playing in the middle of the day. A majority of the time, these youngsters do not need rescue and should be watched and left alone.

Skunks do not hibernate, but during the cold winter months, they enter into a period of greatly decreased activity. Skunks are not highly social animals but will gather in communal winter dens. In one instance, over five hundred skunks were found living under a barn in Nebraska.

There are many theories about when skunks can and cannot spray. Many people have been told and believe that a skunk in a trap cannot spray because it cannot raise its tail. We can personally assure the reader this is not true. Over the years, we have captured full-grown striped skunks in small squirrel traps. These poor creatures couldn't even take a deep breath, much less raise their tail. When a threat is detected, the tail moves to the side and a shot is fired. Even in these cramped quarters, their aim is very good. If you are going to rescue skunks, you are occasionally going to get sprayed.

Another theory we suggest you dismiss is the idea that a skunk won't spray if you hold its tail down. True, it won't be able to spray out to its normal effective range, but it sure can release enough spray onto its tail to ruin the day for even the best tail holder. Others will tell you they won't spray if their hind feet are off the ground. We suggest you don't believe this. It has been our experience that when a skunk feels the need to spray, it's going to spray.

The only real predators of the skunk are raptors, especially the great horned owl, and humans. Skunks are slow movers and are easy to spot. Although their spray is enough to keep other predators away, most raptors do not have a sense of smell.

When a skunk receives a serious injury, it will almost always spray. If handled properly, however, skunks can be approached and rescued without spraying a second time. Because they don't easily panic, a slow and gentle approach can result in a much more pleasant rescue.

The skunk is a natural digger and has long claws. They are long and wicked looking, but they are not sharp and pose little threat to the rescuer.

The skunk will almost always give warnings before it sprays. It will raise its tail straight up and will stomp its front feet. If the threat continues, it will duck its head, turn its business end toward the threat, and spray. When approaching a skunk, move slowly. If the skunk warns you of an impending squirt, stop and let it calm down. Continue to move and stop until you are close enough to place a towel over the animal. It will immediately try to walk out from under the towel, so you will need to hold the towel in place. Gently hold the towel down; don't be rough with it. Unlike the raccoon, skunks will often calm down and submit after covered. If you are gentle and lucky, the skunk may not spray. Another method to cover your skunk is to use two long ropes tied to your towel. This method requires two people. Keeping a safe distance, walk on either side of the animal and pull the towel slowly toward the animal. Very slowly walk the towel over the animal and then approach. If it does spray, at least the spray will be absorbed in the towel rather than your shirt.

Wrap the towel around the skunk and gently place it into a carrier. Keep it wrapped in the towel. Inasmuch as skunks are slow-moving animals, they are susceptible to being hit by cars. This often results in litters of baby skunks left without a mother. It would be difficult to find a baby animal more adorable than a baby skunk. Because they are so cute and cuddly, rescuers of these litters often want to keep and raise them. This is a bad idea.

Wild animals are not pets. They belong in the wild. Baby wild animals are often very fragile and need specialized care. Improper formula, incorrect feeding schedules, and other errors in procedures will often result in the animal's death. It should also be noted that in some states, it is illegal to possess any wild animal. When a nest of baby skunks is found, they should be turned over to a licensed rehabilitator.

Skunks at a Glance

Range	All parts of the United States.
Description	Black body with white stripe on face and forehead. Broad white stripe split in V shape on nape.
Size	12 to 18 inches long [not including tail], 3 to 12 pounds.
Diet	Insects, rodents, berries, fruits, dog food, garbage, birds, bird eggs, and worms.
Reproduction	One litter per year in the spring. Gestation 62 days. Litter size 4 to 8. Weaned at 50 days.
Life Span	8 to 12 years.

After-Rescue Clean-up

When you get sprayed, there are several remedies to deodorize skunk spray.

Everyone seems to have a remedy for getting skunk spray off, especially off the family dog. Everything from vinegar washes to tomato juice baths have been used. Some of these will partially work, but let's face it—the stuff is hard to get off.

There are several remedies that will deodorize skunk spray. One type is mechanical. Ozone and ion generators are commercially available machines that generate ozone and negatively charged ions. The ozone oxidizes the spray chemicals and neutralizes them. The ions react with the positive particles in the skunk spray and neutralize them. These machines are expensive but work well.

In addition, these mixtures are very effective at removing the odor. Mix one quart of 3 percent hydrogen peroxide with one-half cup of baking soda and one tablespoon of Dawn® dish soap. Apply this mixture to affected area and let stand for ten to fifteen minutes. Wash with warm, soapy water followed by a rinse with warm, clean water. Repeat if necessary. This mixture may cause some discoloration to your animal's fur. Do not store leftover mix in a closed container because the chemical reaction may cause excessive pressure and bursting of the container.

Tomato juice is a remedy that is well known and been used for years. Soaking in tomato juice will help, but the chemicals in the juice that react with the skunk spray are in very small concentrations. To achieve the desired result, a very large volume of juice is required. The other methods listed above are much better.

Rabbits

Rabbits are very interesting little creatures. Among homeowners, this nuisance animal probably has more friends than any other pest. They are small, cute, and quiet, and they do not interfere with pets or garbage.

Gardeners are the obvious exception to wanting to live around rabbits. Rabbits will eat almost anything green and can eat a lot of it. Of all the rabbit species, the cottontail seems to be the one needing the most help.

Rabbits are also unique in the animal world because their nests are so easily uncovered by a homeowner and predators. People who have pets or children and have cottontails on their property have a high likelihood of finding a nest of babies.

There are three species of cottontail rabbits in the Northern Hemisphere.

- Mountain cottontails (*Sylvilagus nuttalli*)
- Desert cottontails (*Sylvilagus audubonii*)
- Eastern cottontails (*Sylvilagus floridanus*)

These animals can be found throughout the United States and in much of Canada and Mexico. Rabbits tend to be social animals and group together in colonies. Their presence is determined by food availability. Cottontails thrive on most plant life and prefer grasses, hay, alfalfa, and garden goodies. They feed from daylight to dusk and are especially active when the grass has dew on it. This is how they get most of their water. Cottontails breed year round and can have up to twelve

litters per year. The gestation period is twenty-eight days, and the litters range in size from two to twelve. If all the babies, or kits, survived and the population went entirely unchecked, a single mating pair would have a family of over three hundred and fifty thousand in five years. In fact, only about two kits from every litter will ever see adulthood, and the average lifespan for these rabbits is less than two years.

Cottontails at a Glance

Natural Foods	Grasses, roots, vegetables, and leafy plant matter.
Territorial	No. They live in colonies.
Range	10 to 100 acres.
Hibernation	No. Feeds year round.
Reproduction	Year round, up to 12 litters per year.
Gestation Period	28 to 30 days.
Litter Size	3 to 11 kits.
Leave the nest	3 to 5 weeks.
Predators	Almost all carnivores and raptors prey on rabbits.
Life Span	Less than 2 years.

A cottontail's body temperature is between 101 and 103°F. Cottontails regulate their body temperature by expelling heat from the small blood vessels in their ears. They have keen senses of hearing, smell, and taste, but their eyesight is only marginal. Rabbits are good swimmers and diggers and can climb trees, although they seldom do. They have a very high metabolic rate, about three times that of dogs and cats.

Unlike some other rabbit species that dig and live in burrows, cottontails make their nest by carving out a bowl-shaped depression in the ground in high grass or brush. The depression is about seven inches in diameter and up to five inches deep. It is lined with dried leaves and grass and belly hair from Mom.

The kits are born hairless, eyes closed, and totally helpless. Their ears are not fully developed at birth and are pulled back against the side of their head. The kits develop very quickly. At ten days, their eyes and ears open. They begin the weaning process as soon as their eyes open, and they start chewing on the nesting material.

The mother rabbit does not stay with the nest; therefore, she does not attract predators. She returns to the nest and nurses the babies twice a day, usually at dawn and dusk. The remainder of the time, she remains hidden in the brush, always keeping a watchful eye on her young. Her primary defense, like that of all cottontails, is to stand completely motionless until danger moves away.

In preparing for feeding, the mother approaches the nest, removes the cover material, and lies down on the litter. The babies don't suckle but rather open their mouths and throats so the milk runs directly

Baby rabbits develop quickly. At ten days, these rabbits are starting to open their eyes.

into their stomach. After feeding, mom stimulates each baby to induce defecation and urination. She does this by licking the baby's anus. This entire process usually takes less than a couple minutes. This behavior trait is responsible for many rescues that are not necessary. Baby rabbits in distress will cry, squeak, and crawl out of the nest. If they are quiet and sleeping, they are not in need. Watch the nest until they either get distressed or until mom shows up to feed before disturbing the nest. If a homeowner picks up the babies, try to put them back in the nest and wait to see what mom does. It has been our experience that mom will notice the nest has been disturbed and will move her babies to another location. Baby rabbits are very hard to keep alive during rehabilitation, and they are therefore much better off if left with mom. At between three and five weeks, the kits will be out of the nest and feeding on their own. They are sexually mature and ready to mate between four and six months of age.

Cottontail rabbits cannot be domesticated and do not make good pets. Attempts to keep a wild rabbit in captivity will almost certainly lead to the animal's death.

Most gardeners who have bunnies want to get rid of them. This is understandable because rabbits eat from dawn to dusk and cause a lot of damage. A lot of rescues are the result of homeowners trying to eliminate the animal.

Rabbits are notorious for wandering into the path of dogs, cats, and motor vehicles. There are several species of rabbits, but their defensive behaviors are similar. They are fragile animals and are highly susceptible to stress. They require careful and gentle handling.

Rabbits typically defend themselves with their teeth and hind legs. Their teeth are designed to be very efficient at harvesting vegetation and are very sharp. A rabbit can inflict a deep wound to a human hand. A rabbit bite is usually a slicing wound rather than a puncture wound. Rabbits also have very powerful back feet, which they use in their defense. Their violent kicking cannot only inflict deep scratches but can also break the hold of almost any predator. As soon as it is picked up, it will begin kicking violently in an attempt to escape.

When approaching a rabbit, approach slowly. A spooked bunny can move very fast. Get the rabbit under a towel and hold it down until it calms. If it requires first-aid, be sure to keep the head covered. To pick it up, wrap it completely in the towel and hold its hind feet together. Because of its sensitivity to stress, it is best to transfer the animal as quickly as possible with a minimum of handling. Once it is in the carrier, use another towel to cover the carrier so the rabbit is in a dark, quiet environment. Apply heat as soon as possible.

If the animal needs immediate medical care and you get assistance from your local vet, remind the vet that amoxicillin, penicillin, and clavamox will kill rabbits. Use Bactrim, Albon, or some other sulfa-based antibiotic.

If you are rescuing baby bunnies, pick them up with their natural nesting material if possible. Place the bunnies in a carrier with their nesting material and get them on heat immediately. Do not try to feed them or give them fluids. Rabbits have very special dietary requirements and are highly susceptible to bacterial infections.

Coyotes

Family : Canidae Order : Carnivora

Mankind has been trying to control coyotes for hundreds of years with little success. They are one of the few wild animals that survive better around humans than they do in undeveloped areas. As we gobble up wildlife habitat, many species suffer. The coyote, however, increases in population and thrives.

Coyotes at a Glance

Natural Foods	Fruits, rodents, rabbits, birds, bird seeds, sheep, carrion, bugs, an occasional deer, pets, pet foods, and human garbage.

Territory	Less than one square mile up to as many as 40 square miles.
Size	20 to 35 pounds
Common Diseases	Distemper, rabies, hepatitis, parvovirus, and mange [caused by mites}.
Hibernation	No.
Reproduction	Once a year in the spring. Litter size 4-8.
Gestation Period	62 days.
Predators	Humans, mountain lions, and wolves.

In the 1800s, coyotes were found only in the western United States. In the 1940s, coyotes were spotted in New York and are now in every state and most of Central and North America. Coyotes are adaptable animals that live in a wide variety of climates and conditions and adapt to man as well as any animal known. In fact, experts report that coyotes survive better around humans' neighborhoods than they do in undeveloped areas. They are known for killing sheep and raiding chicken coops and rabbit hutches, which makes them unpopular with farmers and ranchers. When man expands into the coyotes' territory, the coyotes kill house pets, which adds to their unpopular status.

There is evidence that coyotes have been around for millions of years. Coyotes are omnivorous and can adapt their feeding habits to any food that is available. They usually feed at night or in early morning. Their typical diet consists of fruits, rodents, rabbits, birds, sheep, carrion, bugs, an occasional deer, pets, pet food, and human garbage. Coyotes compete for food with foxes and bobcats, which can have an effect on their population. They are important in the control of rodents, preying on rats, mice, gophers, and squirrels. Coyotes weigh twenty to thirty-five pounds, can run up to forty-five miles per hour for short distances, and are formidable hunters. They are normally solitary hunters but, on occasion, hunt in pairs and less often in packs. The coyote has an established feeding route that it uses daily and ranges up to one hundred miles, depending on the availability of food and coyote population. Their normal feeding range is from less than one square mile up to as many as forty square miles.

The coyotes mate between January and March, have a gestation period of sixty-two days, and produce litters of four to eight pups. The mother nurses the pups in the den for about six weeks.

The male stays near the den while the pups are nursing. The pups will be independent by eight months and can breed at one year of age but usually wait until their second year.

In the winter, families usually disperse. Some pups stay in the territory and become transients, and others find new territories and seek out mates to form alpha pairs. Coyotes live from eight to ten years and, like wolves, have been known to mate with domestic dogs. In late winter, the alpha pair mates and begins seeking a den site. They often use the same den each year, but the female may dig a new den or several new ones. Coyotes use a den only when they have their young. The rest of the year, they sleep on open ground in places known as "lays," or they sometimes seek shelter by digging "weather dens."

Once an alpha pair forms, the animals establish a territory that encompasses their regular hunting area. They sometimes tolerate territory trespass by younger animals but normally defend their territory vociferously.

Inside their territory, coyotes have what's called a "core area." This core area is intensely guarded and encompasses about 7 percent of their entire territory. The pair spends about 80 percent of their time in the core area. The den is always located in the core area. Coyotes will not allow intruders in the core area or near the den. While Mom takes up a position in the den to protect it, the male threatens intruders and will fight to the death to protect his territory.

Coyotes are often found in areas of high human population. Humans provide a wealth of food and a variety of shelter options. Many people think coyotes hunt in packs and are extremely dangerous animals. Coyotes know that, in general, people pose little threat, so they become unafraid. Being unafraid and being aggressive are clearly different. It cannot be denied that there have been reports of people, usually children, being injured by coyotes, but it is very rare. Coyotes often stay in pairs or alone in their territory. If the pups stay with the pair to form a pack, the grown offspring become beta animals and seldom reproduce. All animals in a pack are family members.

Studies show that attempts to control and reduce coyote populations have been ongoing for hundreds of years. Trapping, den hunting, helicopter hunting, poisons, and artificial birth control are just a few of the methods tried.

Since the early 1930s, the United States Government has led the attempts to reduce coyote populations. Probably the most widely used government control tool was the Coyote-Getter, also known as the Go-Getter. This device, introduced in the 1930s, used a 38-caliber cartridge loaded with cyanide. The device was placed in a bait source so that when the coyote investigated the bait, the cartridge would fire cyanide into its face and mouth. This device, now known as the M-44, is still available for use on private property. While the federal government developed devices and poisons, many ranchers and individuals began using non-registered poisons like rat poison and anti-freeze. Many of today's coyote rescues are a result of attempted coyote elimination.

In the 1940s, the government and ranching community began using a new poison called Compound 1080. This was a very potent and effective poison, but it was not popular with many ranchers because it also killed domestic animals. An animal exposed to 1080 dies a slow and painful death. Many people familiar with the effects of 1080 labeled it the most inhumane poison ever developed.

In 1972 the federal government banned the use of animal poisons on public property. If rescuing a sick coyote, always keep the possibility of poison in mind.

Even with the massive efforts to try and reduce coyote populations, nothing has had an impact. The U.S. Fish and Wildlife Service reported that, between 1937 and 1981, the agency had killed 3,612,220 coyotes, and in 1982, the coyote population in the U.S. was at an all-time high. Mankind continues to lose the battle.

Nature is one of the best controllers of coyote populations. Pups are born in late winter or early spring, before rodents and other foods become available. Many pups die from starvation, disease, and the elements. On average, only two pups from each litter live to see September. This natural phenomenon is often counteracted by the food provided by humans.

Humans can also cause an increase in coyote populations by killing the alpha animals. In a family pack, the adult offspring, or beta animals, usually do not reproduce, and they sometimes continue to hunt with the alpha pair.

If the alpha female is removed from the pack, the beta animals disperse and begin looking for mates, thus establishing new alpha pairs. Many will establish their territory within the original territory. Studies have shown that when family units break up and new units are formed, the resulting territories are often much smaller than the original territory. The final result is the formation of smaller, closely grouped territories, with more alpha pairs and, ultimately, more pups in the same area.

As mankind continues to move into rural areas, the coyote/human conflicts continue. Efforts are made every day to remove nuisance coyotes and every day more coyotes need rescue.

Coexisting with Coyotes

Coexisting with the coyote is a concept that is slowly starting to emerge in areas where the human and animal habitats significantly overlap. Homeowners in many areas are beginning to landscape to deter pesky creatures and are learning to remove temptation by eliminating food and water from their property. The willingness to coexist with coyotes is moving more slowly than it is with many other wildlife species. This is simply because coyotes have a bad reputation. They are viewed as dangerous and unpredictable. It is difficult to find anyone who really understands or likes this species. Some of the farmers and ranchers (especially sheep ranchers) who have suffered financial losses from coyotes may never accept coyotes as part of their world. To suggest that these ranchers should coexist with coyotes would be like asking a snake to embrace a mongoose.

In addition, many of today's coyote problems occur in residential neighborhoods. Coyotes coming into yards where children play, pets live, and poultry and rabbits roam is a serious concern of homeowners. People expect all roaming wildlife to be controlled the way most municipalities control roaming domestic dogs.

This concept of coexistence is hampered by the lack of education about coyotes. We feel that as rescue people, we have an obligation to try and educate other people about the problems and the animals.

Coyotes, like many species of wild animals, visit homeowners to find food, water, or shelter. In the case of the coyote, once its feeding territory is established, the animal continues visiting the same property until the food, water, or shelter resource is no longer available. Coyotes also aggressively protect and defend their territory so no other coyotes can cash in on all the goodies. For the property owner, this means the coyote that is there today is the same animal that was there last week. This fact about coyote behavior provides the property owner an accurate view of the scope of the problem. If the property owner selects the easy fix (killing the offender), what happens next? The territory becomes vacant, and another alpha mating pair moves in, perhaps this time with a family pack. History has demonstrated that in areas with good habitat, there is a never ending supply of coyotes.

To reduce the problem, remove all available food (including pet food and fallen fruit), water (including bird baths and standing water), and shelter from your property. If a pet is at risk, don't let the pet out alone, or secure the property line to make it very difficult or even painful for the predator to get to the pet. The use of electric fences can help keep roaming animals out. Although coyotes can take livestock at any time of day or night, at night, when attacks are the most common, put livestock in a secure pen impervious to coyotes. Build a large dog run with a screen top, or perhaps get a bigger dog to protect pets. In a report by F. Robert Henderson, Kansas State University, it was stated that installing lights in the feed area greatly reduced the damage to livestock. There are many things that can be done to reduce the threat. The problem probably won't ever be completely eliminated but could perhaps be made tolerable with basic education.

The issue that continues to encourage the outright killing of all coyotes is the fear of being attacked or having children attacked by wandering coyotes. All parents will certainly agree that this is a valid concern. No one will deny that coyote attacks have occurred, but if you look at the entire population of coyotes and the number of human/coyote encounters and couple those data with the actual number of attacks, the probability of a coyote attacking a child is about the same as or less than the probability of that same child being struck by lightning. Coyotes, unless they are sick or starving, are not aggressive toward humans, and they avoid human contact. According to Paul Rezendes in his book, *Tracking & the Art of Seeing:*

> It is worth mentioning that, as of 1989, in all of recorded history, there had been only ten documented cases of human beings having been attacked by coyotes (this includes Canada as well as the United States). In contrast, the Humane Society of the United States reported that from 1979 to 1986, 121 people were killed by domestic dogs in this country, for an average of 15 people per year, and that every year, it estimates, between 1 million and 3 million people are attacked by domestic dogs [pg. 194].

Rescuing Coyotes

With so many coyotes destroyed each year, it is not uncommon to come across a den of abandoned baby coyotes. Humans, cars, predators, and a variety of other factors kill mother coyotes. Most people who find these dens don't have any idea what they should do. People without experience in handling wild animals should not attempt to handle wild coyotes.

Before attempting a rescue, there are several things to keep in mind. In the coyote family structure, the male coyote assists in the care and feeding of the young. This animal will fiercely defend the den. Be sure he is not in the immediate area before approaching a den. If the babies have their eyes open and are climbing in and out of the den, they may be just fine. Coyote pups, like most other mammal babies, clearly let you know when they are in distress. The crying, whining, and overall ruckus they cause is an unmistakable plea for help. If you approach the den and the pups seem content, leave the area, wait for about eight hours, and then return. In eight hours, if the pups aren't agitated, they may be receiving adequate care. If the pups in the den are creating a scene, especially if their eyes are closed, they probably need to be rescued and rehabilitated. You can tell if their eyes are still closed because if they are closed, the pups will not be moving away from or out of the nest. When you approach a coyote den, be cautious and always keep an eye out for adult coyotes.

Use a towel to cover each pup, and put them in a carrier one at a time. They will growl and bite, but with gloves and a towel, they are not likely to cause any trauma. Baby coyotes can easily imprint (tame down), and therefore human contact should be kept to an absolute minimum.

Rescuing an adult coyote is about the same as rescuing a mad pit bull dog. They are very unpredictable and can turn vicious at any time. If you hoop a coyote, remember it is just a medium-sized dog. It is not a wild monster, as many have been taught to believe. Hoop it, pin it, cover it, treat it if needed, put it in a carrier, and transport. Coyotes can carry rabies and inflict serious bites, so wear protective clothing.

There are licensed rehab facilities in every state. It will always be in the best interest of the animal if it can receive the proper care from trained staff. In many states, it is illegal to possess any wild animal regardless of the circumstances unless you are licensed to do so. Coyotes in captivity also need a series of vaccinations and regular veterinary checkups. They do not make good pets.

Squirrels

Squirrels are one of nature's favorite cat toys and often find themselves in trouble.

There are many of species of squirrels and chipmunks throughout the United States. The information below is non-specific to any one species. For rescue purposes, we have divided squirrels into two broad categories: ground or rock squirrels and tree squirrels.

Ground or Rock Squirrels

Ground squirrels are nearly always found in rocky areas such as cliffs, canyon walls, rock piles, pockets under highways, and woodpiles. Although typical ground squirrels in most respects, they can climb trees nearly as well as tree squirrels. Occasionally, they den in trees. The usual den, however, is a burrow dug under a rock or directly into the ground. They are diurnal and most active in early morning and late afternoon. They feed on a variety of plant materials, including acorns, pine nuts, walnuts, seeds of mesquite, cactus, berries of currant and juniper, and all sorts of cultivated fruits and vegetables. Insects also contribute to their diet, especially grasshoppers, crickets, and caterpillars. They are fond of flesh and are known to catch and eat lizards and small birds.

They store quantities of food for winter use which, coupled with the fact that they are seen occasionally in winter on mild days, suggests that hibernation is incomplete and only for short periods. Rock squirrels mate from March through July, which suggests that two litters may be produced each year. Litter size ranges from two to five. Young squirrels first appear above ground at six to eight weeks of age. If something happens to the mother, the babies will leave the nest, searching for food. In the early weeks of development, they will still have their eyes closed and be totally helpless. These wandering babies account for many squirrel rescues. If you find a baby wandering with its eyes closed, it needs help. Although at the time there may only be one, you can be assured more will follow. Sometimes they wander out days apart from each

other, so you must continue to revisit the rescue site on a daily basis until you have them all. Babies that have their eyes open are usually able to take care of themselves, but they are vulnerable to predators.

Squirrels at a Glance

	Ground or Rock Squirrel	Tree Squirrel
Nest Sites	Underground, under rocks, cliffs canyon walls	Hollow trees
Primary Diet	Acorns, pine nuts, walnuts, seeds of mesquite, cactus, berries of currant and juniper, and all sorts of cultivated fruits and vegetables	Plant material, seeds, berries, acorns
Hibernate	Semi	No
Mating	March - July	March - June
Litter Size	Two - Five	Two - Five

Tree Squirrels

As the name implies, tree squirrels spend a majority of their time in trees. They prefer nesting in hollow trees but will sometimes build their own nesting location from forest debris. The tree squirrel's diet consists mostly of plant material, seeds, acorns, and berries. They will, however, also munch on insects, bird's eggs, and small birds. They store up food for the winter and store it in abandoned animal burrows, in hollow logs, and in abandoned automobiles.

Tree squirrels mate from early winter to late spring. One litter of two to four young are produced from March to June. They may have a second litter in late summer. The young will venture out of the nest at between six to eight weeks. At about three months, they will be on their own. Unlike the ground squirrel, the babies are relatively safe and secure in their nest. Typically, if something happens to Mom, they will usually perish in the nest. There are far fewer baby tree squirrels needing rescue. The average lifespan of a tree squirrel is three to five years. Tree squirrels do not hibernate and are most active at dawn and dusk.

Squirrels are the most likely animals to die from stress. Some squirrels will die from stress simply by being caged. Like the rabbit, they are also very susceptible to bacterial infections. They will usually die within hours of being subjected to a cat bite if they don't get treatment. It is very important to keep these animals warm and housed in a dark, quiet place.

Squirrels have razor-sharp teeth. It is not uncommon for them to bite through a leather glove. This is their primary defense, and they can inflict very deep and painful wounds. When rescuing a squirrel, cover it and stay away from its teeth. They can move incredibly fast and can also jump several feet into the air. Never try to pick up a squirrel by the tail because its body is very flexible, and it can have its teeth into your hand before you know what happened.

Most squirrel rescues are the result of being hit by cars or being abandoned after Mom is killed. Squirrels are also very close to the bottom of the food chain, which results in a high mortality rate. Squirrels seem to be more robust than rabbits when it comes to rescue. Although they stress easily, they seem to respond much better to treatment and rehabilitation.

Although squirrels seldom carry rabies, they are a primary carrier of plague-infected fleas. Before handling a squirrel, dust it with flea powder. Never pick up a squirrel without gloves, and carefully watch for fleas on your clothes and equipment.

Baby squirrels are less sensitive to stress than adults are. They are easy to rescue and easy to care for. The only real concern in rescuing babies is to be sure to keep them warm and never let them get wet.

Bears, Mountain Lions, Adult Deer, and Elk

You should never attempt to rescue these animals unless you have someone to help and the proper equipment. These animals cannot easily be captured or held down without the help of a tranquilizing agent. These animals are so powerful that a single swipe from their hooves or claws can cause very serious injury. Tranquilizers are highly controlled substances and should only be handled and administered by a trained and licensed rehabber or a veterinarian. Using these drugs should be the last resort and should not be administered to animals that are still mobile. There is a several-minute time lapse between administration and effect. The animal will likely run away before the drugs work, and the rescue will be a failure. Two people can usually capture baby and juvenile animals using a hoop. Hoop these animals around the neck, and use the pole to keep the animal away from you and your partner. Once hooped, do whatever you can to get the animal off its feet. Cover

Don't attempt large animal rescues without help.

its head with a towel, and let it calm down. Keep on top of the animal, and don't allow it to fight its way loose. Tranquilizers will still be required, but they can be controlled while waiting for the drugs to take effect. Deer and elk are also very susceptible to stress. Try to keep them calm, and don't handle them any more than is required.

Virginia Opossums

Opossums inhabit most of the United States. Opossums are nocturnal and live in forests, grasslands, agricultural lands, and suburban areas. They thrive around humans. They will eat almost anything available including bird's eggs, chickens, rats, earthworms, insects, snakes, grass, fruit, pet food, and garbage. Carrion (animals that are already dead) is also a favorite food. Dens for daytime use include just about anything that provides shelter.

The Virginia opossum is the only North American marsupial. Like other marsupials, such as kangaroos and koalas, opossums give birth to highly underdeveloped young, which then climb along the female's

belly to a pouch, where they develop. Within the pouch, the young attach themselves to one of thirteen milk-providing nipples and do not let go for about sixty days. The average litter size is seven to ten. Opossums usually have two litters per year during the period from January to July.

Baby opossums leave the pouch in about six to eight weeks. They spend the next several weeks riding on Mom's back.

Opossums have a short lifespan, only living from two to three years.

When threatened, opossums can be very vicious animals. They are also highly protective of their young. Sometimes when threatened by a potentially dangerous animal or person, opossums will growl and hiss as a warning. If the intruder does not back away, the opossum will suddenly fall over, open its mouth, and play dead (playing possum). Many predators do not eat dead animals, so they may leave the opossum alone if it appears to be dead.

Knowing this behavior trait can be very valuable to the rescuer. You should always approach an injured and "dead-looking" opossum very carefully. Although probably not as dangerous as a raccoon, the opossum can be a very vicious and dangerous animal.

These animals are slow moving and are susceptible to vehicle collisions. Cars account for many deaths for this species in urbanizing areas. They are also prey to bobcats, coyotes, and domestic dogs.

The most common opossum rescues occur when Mom gets killed and the babies are left behind. Baby and juvenile opossums can usually be handled with just gloves. Adults usually require the use of a hoop. Because of their cone-shaped heads and pointed noses, you will need to get the hoop around at least one leg. It is still a good idea to cover the animal with a towel to reduce stress.

Rescuing opossums is not that difficult because they are smaller animals; therefore, they are easier to handle. The same rescue techniques used on the raccoon should be used on the opossum.

Opossums at a Glance

Habitat	forests, grasslands, agricultural lands, and suburban areas
Diet	bird eggs, chickens, rats, earthworms, insects, snakes, grass, fruit, pet food, garbage, and carrion
Mating	January to July
Litter Size	Seven to Ten
Nocturnal	Yes
Life Span	Two to three years

Porcupines

Porcupines are rodents with a coat of sharp quills (present at birth), which defend them from predators. They occupy a wide range of habitats throughout the world. Porcupines live in forests, deserts, and grasslands. Some live in trees, and others live on the ground. Porcupines are always in search of salt. For homeowners, this can be an issue because porcupines will eat anything with salt. This includes paints, tool handles, shoes, clothing, and wood products. Porcupines are also attracted to roads where salt is used to melt ice and snow. This accounts for a large number of animals needing rescue. Porcupines are strict herbivores, feeding almost entirely on trees. But they also eat ends of branches to get to the leaves, buds, and nuts. Porcupines will eat any species of trees found within their range. Breeding occurs in the fall and is followed by a gestation period of about two hundred ten days. The young are born in the spring. There is almost always only one baby. It is born with soft quills that will harden within hours of birth.

The porcupine is a relatively easy animal to rescue but a little tricky to transport. Porcupines are slow-moving, docile creatures that

seldom become aggressive. They do have very sharp teeth and are capable of inflicting nasty bites. The biggest problem with the porcupine is its massive set of quills. Hook-shaped on the tip, these quills hurt as much coming out as they do going in. If a quill is not removed, it will work its way into the body and could cause death. Folklore tells us the porcupine can shoot its quills by flicking its tail. This is not true. They can, however, cause a lot of pain if they slap you with the tail. This is their main defense. They don't outwardly panic and don't stress as much as most other animals. Like the skunk, they have few predators and don't get terribly excited when approached. This is one animal that shouldn't be covered with a towel when captured because it is difficult to handle a porcupine if you can't see the quills. A towel won't even slow down a quill. Porcupines don't have quills on their chest or underside. This is useful information for anyone trying to put a porcupine in a carrier. Rather than picking the animal up to get it in the carrier, it is much easier to bring the carrier to the animal and scoot it in. Because the quills point from the front of the animal to the back, it is better if you can get the creature to go in the carrier tail first. This will make it much easier to get the animal out later. If field first-aid is required during rescue, you can use an old window screen to hold the animal down. Place the screen wire over the animal so the quills are pressed up against the animal's body. This will hold it still, allowing you to work with it.

Porcupines at a Glance

Habitat	Forests, deserts, and grasslands
Diet	Salt, trees
Breeding	Fall
Gestation	210 days
Litter size	One

Chapter Seven

What to Expect

A majority of rescues are routine and fairly straightforward. Injured animals are usually easy to capture. When caution and common sense prevail, the capture is often uneventful. On occasion, however, the capture can be a unique experience and sometimes even presents a challenge. The following captures are described here to give you an idea of how some rescues can be a little different.

Animals sometimes get themselves stuck in unusual places, and there are a lot of unusual places to choose from.

A call came in to rescue a gray fox that was supposedly caught in a fence. There are many types of fences and several ways to get caught in each. We weren't sure what to expect. This fox, for whatever reason, had pushed its head through an opening in a chainlink fence. It was caught around the neck and was frantic to get loose. It was rolling over and over, causing serious damage to its neck. Needless to say, approaching the animal was easy. As we got close to the fox, it began to panic and caused additional damage to its neck. We placed a towel over its body and held it down until it stopped spinning and pulling. Because of the fence, we were unable to cover its head and eyes. There was no way to remove the animal from the fence, so we decided to cut the wire. Luckily, I knew I was rescuing from a fence, so I grabbed a pair of heavy cutters on the way out. We had been unable to contact the homeowner, so we knew we were facing potential liability for the fence. We cut the fence in three places, removed the fox, and transported it to a rehab facility.

As soon as the animal was secure, we returned to the fence and contacted the owner. We explained what had transpired and explained the need to cut the fence. We assured the homeowner that we would repair the fence. In this particular case, the homeowner was grateful that we rescued the fox and agreed that mending the fence was adequate.

Anytime you rescue an animal on private property, you must remember that you are responsible for any damage to that property. Always try and get permission before entering private property.

We have had several rescues where animals were caught in wire and nets. Garden netting really takes a toll on little creatures and birds. Over the years, we have learned the best method of releasing an animal that is entangled in a net, is to cut the net. You can spend hours trying to untangle animals, causing them unnecessary stress and usually getting bit in the process.

The most difficult rescues involve trying to capture animals that still have full mobility. We were called to rescue a gray fox that had been spotted traveling with a steel leg trap on one of its hind legs. By the time we arrived, the fox had disappeared into a canyon. We spent several hours trying to locate it but were unsuccessful. A few days later, we got several inches of snow and decided to try again using the newly fallen snow to show its tracks. We found the tracks and began the chase. We were carrying a net gun, hoping to get close enough to use it. Eight hours later, we were still on the track. We had seen the animal several times but could not get close. It was bleeding from its wound, and the bleeding was getting worse as the hours passed. We were sure we would eventually run it down. The chase finally ended when my wife slipped on some ice and fell off a cliff, severely breaking her leg.

A couple days later, we set out live traps and were able to capture several foxes but not the injured one. The injured fox was never spotted again.

When called to rescue an animal, especially an animal that is still mobile, you must realize the probability of capture is small. Three-legged animals, even those carrying three pounds of hardware, can still outrun a human. A short chase to see if the animal will hold up is advisable, but if it keeps running, you will probably have a better chance of capture with live traps.

In most cases, if the animal is gone when you arrive, chances are that the chase is over before it even begins.

Animals, especially raccoons, are notorious for falling into things they can't get out of. Garbage dumpsters can be traps for unwary raccoons. You are likely to get many calls to extract critters from dumpsters. The easiest way to get an animal out is to give it a ramp so it can climb out on its own. A rope with knots in it, a piece of carpet, or a stack of cardboard boxes will usually do the trick. If it won't climb out on its own, you may have to hoop it and lift it to freedom.

Sometimes it is a little more difficult.

We got a call to rescue a skunk that fell into a dry well. The well was about fifteen feet deep and about four feet by four feet. The walls were concrete. The only thing I knew for sure about this rescue was that I was not going into that hole.

I considered trying to lasso it with a rope but became concerned about what might happen when a roped skunk came to the surface. Instead, I got a skunk trap and tied a rope to the handle of the trap. I set the trap and lowered it to the bottom of the well. The skunk saw it, of course, but didn't seem to care. I carefully started moving the trap using the rope. I maneuvered the trap toward the skunk until it was against one of the walls. As the skunk moved, so did the trap, trying to keep it in a corner. Finally the skunk entered the trap. I lifted the trap about three inches off the ground and let it fall. The impact of the falling trap set it off. I now had a skunk in a skunk trap. I slowly raised the skunk to the surface and relocated it. The skunk never sprayed.

An injured animal will sometimes seek refuge in a tree. If it's a tall tree, this can cause problems for the rescuer. Going up a tree after a wild animal can be very dangerous and is not advised. A simple trick that may get an animal out of tree is to spray it with a garden hose. Don't hit it with a direct spray, but aim the water stream just above the animal. Most animals don't like to get wet and will retreat down the tree. If the animal is in a tree where water won't work, you will have to go to Plan B. We were attempting to rescue a baby raccoon out of a tree that was being cut down. The baby was about sixty feet up and far out on a very small, flimsy limb. Using a hose would have washed it out of the tree. Instead of a hose, we acquired a long-handled pruning sheer and sent the homeowner up as high as he could get on a scaffold he was using. Then with the help of a few neighbors, we got a blanket, stretched it out under the coon, and had the homeowner clip the branch the raccoon was on. Down he came, branch and all. We successfully caught the little coon in the blanket without any injuries.

We were called to pick up a baby skunk that had been rescued by a homeowner. A car had hit the mother and left the baby stranded on the roadway. When we picked it up, a young girl told how she had seen another skunk go under a shed days earlier. Using flashlights, we looked under the shed but didn't find any skunks. Because of its size, we couldn't move it. We set out to round up neighbors and any strong backs we could find to provide enough horsepower to tilt the large building. We managed to get the front of the building off the ground far enough to allow someone to crawl under it. Lo and behold, we found eight more very hungry little skunks.

Many animals (except porcupines) have babies in litters. When rescuing a baby, ask questions and look around the property for additional babies. If there is one, there are usually more.

A very popular place for injured and scared animals to hide is in the engine compartment of vehicles. We have had several requests to extract animals from cars. Once an animal (especially a raccoon) lodges itself in an engine and tangles itself in all the wires, hoses, and manifolds, it is virtually impossible to get it out. It can hold on to a menagerie of things in there. Pulling and tugging on it is a good way to get bit and to injure the animal. The best way to get it out is to back away and let it come out on its own. Once it feels the threat is gone, it will come out. If it doesn't, you will need to contact a veterinarian and sedate the animal.

Smaller animals (mice and chipmunks) like to nest in car heaters or somewhere up under the dashboard. Anyone who has ever tried to work under a dashboard can understand the challenge of getting an animal out. The best way to remove these animals is to set live traps in the car and get them when they come out to eat. This usually doesn't take long. If you are called to remove an animal that has been living in the car heater, you should advise the car owner that rodents build their nests with flammable materials and they should take their car to a mechanic and have the nest cleaned out. We have seen several car fires as a result of chipmunk nests.

During the early summer, you will probably get calls to remove raccoons from chimneys. The animal is usually heard rather than seen, or an adult raccoon is spotted coming out of the chimney. You can be sure that if you have an adult raccoon in a chimney in the springtime, you will also have a nest of babies. It is crucial that you get the babies out as soon as possible because the creosote and soot in a chimney is highly corrosive and will severely burn the babies. There are many fireplace designs, but you will almost always find the babies on or very near the damper. To get them out, you will have to get into the fireplace. This is a very dirty job, but someone has to do it. Before reaching into the chimney to locate babies, you must be sure Momma has left. Sometimes she will go up the chimney a few yards to get out of your reach, but she can still see her babies. Momma coons are very protective and will not want you bothering the babies. To be sure she is all the way out, you can run a short (10 ft.) length of garden hose up the chimney and rattle it around. You will need someone to watch the chimney opening to tell you when she has left and to warn you if she comes back. After she is gone, reach up into the chimney from the firebox and try to locate the babies. Scratch the sides of the chimney to get the babies to make noise; this will help you locate them. Once located, pull them out one at a time. Baby raccoons still in the nest don't have teeth yet, so they won't hurt you. If you can't reach them, you will have to start looking for tools to help. A large aquarium net is a possibility. Tongs will sometimes help, but don't squeeze the babies too hard. Remember, you cannot give up; the babies will die, or at best, be blinded.

Make sure you have all the babies. Check very carefully. Sometimes one baby will crawl away from the litter.

If you are called to rescue a nest of baby raccoons that are not in a hazardous environment, you should first try to get the mother to move the nest. Baby animals are always much better off if left with their mother.

Raccoons like dark, quiet, warm, and secure locations for their nest. This is why they often select places in attics, basements, garages, and fireplaces. To get the mother to give up her selected nesting area and move the babies, you can disrupt her happy home by introducing bright lights, noise in the form of a radio, and perhaps even a house fan blowing in their direction. Look in on the nest often and make lots of noise. Spray air fresheners into the nest area. If you can make the nesting area unpleasant for the mother raccoon, she will find a new place to live and move the kits.

We had received reports of a young raccoon with a limp frequenting a homeowner's patio. It came around every day at the same time, so we were reasonably sure we could trap it. It only took one night and we had it, or so we thought. We were using an enclosed skunk trap that was only open on one end. When we went to retrieve the coon, I picked up the trap and looked in to see the raccoon. To my surprise, I was eyeball to eyeball with a fat little skunk. It didn't take long for me to look at him.

When trapping an animal, approach your trap with caution. With so many animals out there, you can't always be sure you trapped what you think you did.

One evening we got a call from a hysterical homeowner, stating there was a huge wild animal in her house. She didn't know what kind of animal. All she knew was it was a big one.

When we arrived at the house, we found a lady and her young daughter standing in the yard. They were terrified. I tried to get more information, but all she could tell me was it was wild and big. I assumed it must be a dragon.

I went into the house and conducted a search, but I couldn't find the dragon. I called the lady to the doorway and asked her where she had last seen it. She looked in the house, pointed at the sofa, and told me it was right over there. I found the dragon and had a problem not laughing when I told her the dragon was a folded-up elk hide. Her husband had brought it down from the attic and not told her. Rather than transporting it, I just let it sleep.

On another rescue, we spent over an hour searching for a snake in a garage that ended up being a wrinkle in a rug. We have also been called twice to pick up a little animal that was chirping and must be in trouble. In both cases, we found our little creatures to be smoke detectors with a low battery.

Chapter Eight

Lessons Learned

When we began rescuing animals about twenty years ago, there were no training programs available and no one to give us advice. We learned by doing. Through our learning process, some of our knowledge came from "The School of Hard Knocks." The following war stories describe some of the lessons we learned. We hope our readers can learn from our experiences.

A homeowner requested our services to rescue what he described as a very sick raccoon in his backyard. When we arrived, we located an adult raccoon lying under a bush. It wasn't moving and allowed us to approach with little objection. An exam of the animal didn't reveal any obvious injuries. We pulled the raccoon out from under the bush and covered it with a towel. It was strong enough to raise its head but made no attempt to resist us. We placed the animal in a box trap for transport to the vet. When we arrived, the vet asked us to remove it from the trap for examination. I was wearing double-bite gloves and felt confident that I was adequately protected. The raccoon had been placed into the trap headfirst. I grabbed its hips and attempted to pull it out. The raccoon took hold of the trap's trip plate with its front feet and wouldn't budge. I continued to pull with one hand and reached into the trap with the other to force the coon to give up its grip. As soon as I grabbed the animal on the chest, it turned and got a hold on my thumb. The gloves were adequate to keep the teeth from penetrating, but the strong jaws started crushing bone. It broke two bones in my thumb. I could both hear and feel the bones breaking.

Lesson Learned

From this event, we learned three valuable lessons. First, no matter how incapacitated an animal appears, never let your guard down. Wild animals are unpredictable and should always be considered a threat. Secondly don't put your hands into a confined space with a wild animal, especially when you can't see its head. Finally, don't allow yourself to become overconfident about the effectiveness of your personal protective equipment.

We had trapped a nuisance raccoon and were preparing to pick it up. The trap had been set on a sawhorse to avoid catching skunks. I started to remove the wires that held the trap in place. I was on my knees next to the trap. I was only wearing lightweight work gloves, so I was watching carefully so the coon didn't bite my hands. Suddenly, without any warning, the raccoon struck out with a front paw. The paw went between the wire mesh and caught me on the lower lip, causing a deep cut. The mesh size on the cage was one-by-two inches, and I don't think the coon ever touched wire.

Lesson Learned

Anytime you are handling an animal or a cage containing an animal, keep your entire body out of the animal's reach. It doesn't care if it gets you on your hand, your leg, or your face. It just wants to get you.

We were relocating a trapped skunk. For us this had become a very routine activity that seldom posed any threat. On this particular day, we had several calls pending and were in a hurry. Our normal procedure for releasing skunks was to get them at least one hundred feet from the truck before letting them go. Because we were in a hurry, we only took the skunk about ten feet from the truck and opened the trap. Very seldom do skunks spray when being released, but on this day, the skunk must have been having a bad day because when he came out of the trap, he turned his business end toward us and let loose. We were quick enough to get out of the way, but the main stream went through our open front window into the cab. The rest of that day was "ugly."

Lesson Learned

Never lose focus on what you are doing. When an activity becomes routine, it is easy to cut corners. If you get in a hurry and cut corners, you may end up paying a price. Handling animals requires care and full-time concentration.

We rescued an adult rock squirrel with an injured leg and were getting ready to transport it. I was wearing single-layer bite gloves. I was holding the animal, and my trapping partner (also my wife) was holding the carrier upright to make it easier for me to drop the animal in. When the squirrel realized it was about to get put into a box, it made one last gallant attempt to wiggle free. As I forced the squirrel through the opening, it reached out and bit my partner on the finger, causing a deep wound. She was not wearing gloves.

Lesson Learned

When handling an animal, you must not only focus on your safety but also the safety of everyone in the immediate area. Wild animals don't care who they bite.

We were transferring a wood rat from a trap to a deep aquarium. We had limited knowledge of the behavior of this species, so we were going to keep it and watch its behavior. We dumped the animal into the aquarium, but before we could get a cover on it, the rat jumped straight up about three feet into the air. My reflex reaction was to reach for the flying creature, which I did. I grabbed the animal in midair, and it proceeded to bite through my thumb.

Lesson Learned

Reaching for the animal was a reflex action that was probably unavoidable. The problem was that I had no idea that wood rats were spring-loaded and could jump like a kangaroo. If you are going to be rescuing a particular species of animal, you should do some research on the species before encountering it in the field. The more you know, the fewer problems you will encounter.

We rescued a raccoon that had wandered out in front of a car. It appeared to have head injuries and also a few superficial wounds on its hips and legs. We admitted the animal into rehabilitation and started treatment for its head injuries. After about ten days, the raccoon was moved to an outdoor habitat so it could get some fresh air and sunshine. It still had a tilted head and was a little lethargic but seemed improved. The outdoor cage we chose was adjoined by another habitat containing five baby raccoons being rehabilitated. After about five days, the babies started dying. They became disorientated, started falling over and developed runny eyes and noses. We lost three babies in four days. On the fourth day, the head injury raccoon also died. A careful examination indicated the adult raccoon had probably been suffering from distemper. The symptoms of distemper and head trauma are very similar. Because a car had hit the animal, we assumed we knew it was head trauma. If it had been suffering from distemper, that probably contributed to its getting hit. When we put it in a cage in close proximity to other raccoons, it probably spread the disease to them. Distemper is a highly contagious disease.

Lesson Learned

Whenever an animal is rescued, it should remain isolated until you are sure it is not diseased. If it shows any symptoms of a disease, it must be assumed to have the disease and it must be isolated. Any equipment that was used to rescue that animal must be cleaned and sanitized before it contacts another animal. Equipment should be soaked in a solution of Clorox® and water in a 1:10 ratio. Do not take chances with diseases.

We were picking up a trapped raccoon for relocation. As we approached the trapped animal, we noticed a juvenile raccoon on a tree trunk about five feet from the trap. It was obvious that we were dealing with a mother and her baby. The baby was about half grown. I was wearing lightweight leather-work gloves. I wanted to keep the two raccoons together, so I reached out and grabbed the baby off the tree. It was a very young animal but it had good teeth, good claws, and a real attitude. I had the baby in my clutches and was ready to put it into a carrier. I wasn't only ready—I was anxious. Unfortunately, I didn't have anything to put the animal in. I instructed my partner to get a carrier from the truck while I continued to wrestle with the coon. By the time the carrier arrived and the raccoon was put away, I had multiple cuts and bites on both arms and hands.

Lesson Learned

Before you approach an animal to rescue it, stop, think, and formulate a plan. In this situation, I should have backed away, sent for a carrier and better bite gloves, and waited until I had the proper equipment. This little raccoon gave no indication that it was going anywhere without Mom. Don't let yourself react carelessly to a situation that suddenly arises. Back off and formulate a plan.

We were preparing to transport a rehabilitated squirrel to a relocation site. We had placed the animal on the truck and were about to leave when some neighbors stopped by to chat. The neighbors had their six-year-old daughter with them. We were all standing behind the truck talking when we heard a scream coming from the little girl. She came running to her mother with blood all over her hand and arm. We learned that she had spotted the squirrel in the truck and wandered over to get a closer look. She wasn't more than five feet from the group of adults. She reached out and put her fingers in the cage to pet the squirrel. She was taken to the hospital and got several stitches.

Lesson Learned

When you have a wild animal in your possession, that animal is your responsibility. If it bites someone, it is your fault. Do not show the animal to people, do not allow anyone near the animal, and never leave the animal unattended.

We hope these lessons will help you avoid making the same mistakes. Accidents will happen, but if you have knowledge and experience, the accidents can hopefully be very minimal.

Every time you rescue an animal, you will learn something about handling animals. If you couple this knowledge with a thorough knowledge of the animals and their habits, you will become a successful rescuer. Rescuing wild animals is at times a little scary, but the rewards are immeasurable. Picking up an animal on the verge of death and then later watching it run wild again is what this business is all about.

We hope this book has been helpful in giving you enough information to get started and have the confidence to do the job.

Chapter Nine

Coexisting

When man appeared on the planet, he was met by a variety of animals. On that day, human/animal conflicts began.

In the beginning, animals served as a resource to help man survive. Animals provided man a food source. In turn, man also became a new food source to some animals. In this dog-eat-dog environment, coexisting was a part of life. As time passed, new cultures evolved and new technologies changed the way animals and humans interacted. Some cultures continued to look toward wild animals as food sources, and some cultures viewed animals as sacred. In most cultures, however, wild animals became much less important as a needed resource. Domestic animals became the primary food sources, and man had learned to protect himself against being eaten by wild animals.

As people continued to expand into rural areas, animals were pushed further and further into over-crowded habitats, forcing them to adapt and live in close proximity to people. People begin to see the animals as pests rather than usable resources. Over the years, we have witnessed horrendous mistreatment and cruelty to animals simply because they were in the wrong place at the wrong time and simply got in the way.

In today's world, the refusal to coexist is becoming a serious problem. As with many of our natural resources, they are in danger of being eliminated.

Before a solution can be found, one must have a thorough understanding of the problem. There are several factors that inhibit coexistence.

People have always and continue to take natural resources for granted. Animals, for instance, have always been around, and people think they will always be here. As the endangered species list continues to grow, we know this is not the case. Some of the most commonly seen wild animals, e.g., raccoons and coyotes, have learned to live around people and seem to be increasing in population. People don't realize, however, that they are depleting many of our other wild animal species by mismanagement and taking away habitats.

To help understand the true nature of the problem, one must identify what it would take to get widespread coexistence. The first step would be to get people to respect and care about wildlife. Unfortunately, there are people who see no value in animals and have absolutely no respect. There are also those who simply do not have any knowledge about animals and have no interest in learning.

There is a subset of the population who see negative financial impact from animals (e.g. farmers and ranchers). This group of individuals is not interested in coexisting with animals and probably never will be. Homeowners, especially those with pets, gardens, bird feeders, ponds, and fruit trees, are also typically opposed to coexistence.

Then there are those people who would like to coexist but don't know how, while others would like to coexist with only certain species. They, unfortunately, also don't know how.

Over the years, we have discussed coexistence with hundreds of people and organizations. We have found a common denominator and what we consider to be the biggest cause of the problem: a lack of education.

Before you can expect to have positive interactions with animals, you must have an understanding of the animals and their impact on the planet.

If an animal is giving you grief and you want to change its behavior, you need to know why the animal is doing what it is doing.

Another major reason for a lack of coexistence is that people are unwilling to compromise. You are not going to negotiate or cut a deal with a pesky animal. Animals will not leave just because you tell them. They need food, water, and shelter. If an animal is causing problems, it is not personal. Animals don't come onto your property to eat your dog food because they don't like you or they are mad at you. They are there because you left them a source of food. They would be just as happy eating dog food from your neighbor's yard. If you offer these things, they will take them. If you are expecting a compromise with the animal, that is not going to happen. Any compromise made will be your compromise.

If you are interested in coexisting with an animal, formulate a coexistence plan.

The first step is to take the time needed to learn as much about the animal as you can. The more you know about the animal, its behavior, and what is driving it, the better your chances of altering its behavior.

An important thing to remember is that all wild animals need food, water, and shelter. If you know that raccoons, for example, eat grubs and worms, this might explain why the raccoons are tearing up your grass every night.

Once you have figured out why the animal is coming around, then start looking for ways to change its behavior. If, for example, a skunk is coming in to get dog food off your porch, you may need to remove the dog food at night. Although this may be more work for you, and it may be inconvenient for both you and the dog, it will keep the skunk off the porch.

You obviously can't bring your yard in at night, so you must find another way. Keep in mind, animals will always go for the food they like best and will go for the food that is easiest to get. A yard pond with feeder fish may be better and easier for the raccoon than worms and grubs.

There are a few things you can do to keep animals away from where you don't want them. The following is a sample of the kinds of things the homeowner can do.

Coexisting with Skunks

Skunks are one of the easiest and at the same time the most difficult animals with which to coexist. If the animal living under a shed never feels threatened and therefore never sprays, the shed owner can be totally oblivious to its presence. If, on the other hand, the skunk feels threatened and sprays, the shed owner will want it gone right now.

Skunks eat lots of insects and small rodents and when not causing a stink, they are often referred to as the farmer's best friend. They are very clean animals.

The easiest way to coexist with the skunk and take advantage of its good nature is to remove anything the animal may perceive as a threat.

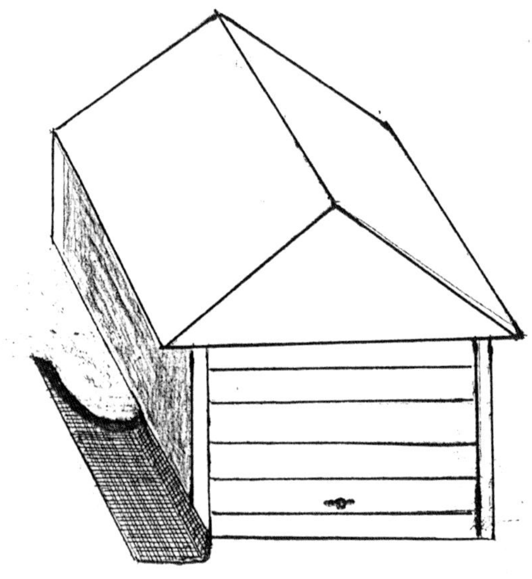

Pets, especially dogs, seem to always excite skunks. To prevent dogs threatening the skunk, fence off the area around the skunk den so the dogs can't get close. Keep the dog food somewhere inaccessible to the skunk. If you want the skunk to den somewhere else, remove from your property what it wants: food, water, or shelter. A chainlink fence around the "no-skunk zone" will usually work. Skunks (except spotted skunks) will not climb a fence. If the skunk digs its way in, attach a twelve- to eighteen-inch-wide piece of wire screen to the fence or building you are trying to protect, bend it into an L shape, and bury the lower portion underground.

Skunks are not nearly as tenacious as many other animals and will usually respond to barriers installed by humans.

Coexisting with skunks is the same as coexisting with any wild nuisance animal. You must keep in mind that these creatures are trying to survive in the habitat we took away from them and that all concessions must be yours.

Coexisting with Rabbits

Wild rabbits can be a real pain for those who are trying to grow a garden. The saying "here today, gone tomorrow" may have originated from a gardener with bunny trouble.

If you plant it, they will come. If you don't protect it, they will stay.

A garden, like the lawn with worms and grubs, cannot be brought in, so you must find a way to protect it. Tempting rabbits away with yummier food is impossible when they have a garden full of young green plans at their disposal, so you must find a way to keep them out. Like all attempts to coexist, this can be a lot of work and costly. Killing bunnies is not the answer because, like the coyote, there is never a shortage of rabbits waiting to move in. The most effective way to protect a lawn or garden is to fence it. Not just any fence, a bunny-proof fence.

Rabbits do not climb fences. If low enough, they will jump fences and if not secured at ground level, they will dig under them.

A bunny-proof fence should be constructed of wire with mesh small enough that even the smallest rabbit cannot fit through. Adult rabbits have been known to chew through chicken wire, so your wire should be stronger. The fence should be three feet high. Once in place,

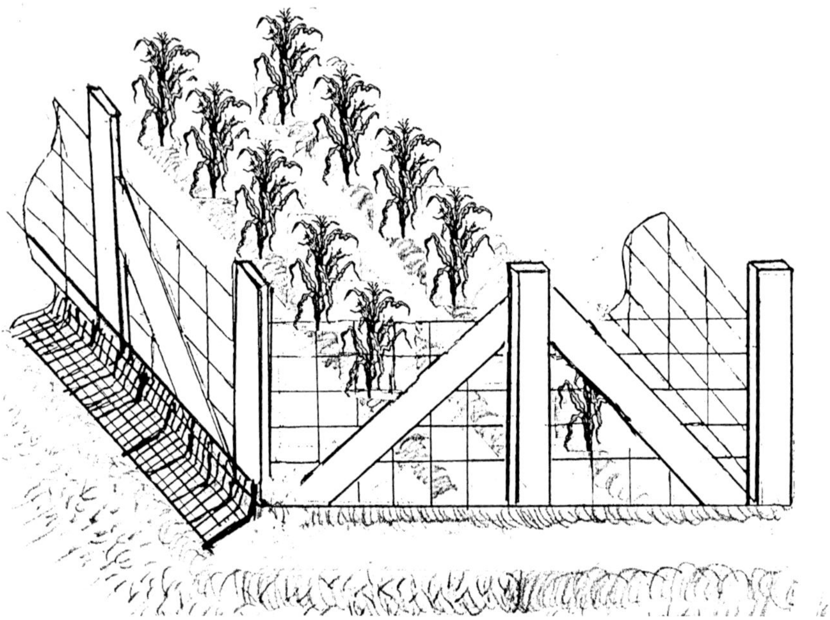

you will need to dig a trench around the fence that is about twelve inches wide and two inches deep. Attach a fourteen-inch strip of wire to the bottom of your fence, let it extend down into your trench, bend it at 90 degrees so the wire covers the entire trench, and backfill the trench.

If the nuisance bunnies dig under your fence, add a twelve- to fourteen-inch section at the bottom of the fence. Bend it into an L shape, attach one edge to the existing fence, and bury the other part about two inches underground, extending out and away from the garden.

If fencing is not an option, you can protect specific plants by making them unpalatable to the cottontail by spraying them with the following compound:

Mix two tablespoons of cayenne pepper with one-fourth cup of cooking oil. Mix until you have a thick paste. Mix the paste with one-half gallon of water and shake well. The paste will not dissolve in the water but will be suspended as red oil drops.

Put the mixture into a spray bottle and spray on plants you want to be off-limits to the rabbit. Continue to shake the solution while applying to keep the paste from settling out. Reapply about once a week. Store leftovers in the refrigerator to keep the oil from souring. This mixture will not hurt plants and is not toxic on vegetables.

Coexisting with Raccoons

Because of their tenacity and apparent comfort around people, raccoons are difficult animals to deter. When working with raccoons, it is good to always remember "the raccoon rule."

The Raccoon Rule

If he can see it, it's his.
If he can't see it, it's his.
If he can smell it, it's his.
If he can't smell it, it's his.
If he can hear it, it's his.
If he can't hear it, it's his.
If he can touch it, it's his.
If he can't touch it, it's his.
If he wants it, it's his
and
If it's yours, it's his.

Raccoons are one of the most common and abundant backyard pests in the United States. They are persistent, curious, and very bold animals. Like most other backyard pests, they are constantly on the lookout for food, water, and shelter. They are also probably the most difficult animals with which to coexist. Many other animals, when deprived of easy food, water, and shelter, will move along and search out new sources. The raccoon will not leave until every inch of the property is inspected and often touched. If you put the dog food inside but accidentally leave a window unlocked, the raccoon is apt to open the window and get the food anyway.

We have had moderate success at controlling the tenacious raccoons by landscaping property to be raccoon friendly. It is never a good idea to artificially feed wild animals because they have specific caloric needs throughout the year. You can, however, provide free-growing or natural foods that might capture the attention of raccoons. Select a portion of the property and put in a pond with fish. Plant some corn and grape plants. Actually develop a habitat heaven for the animals. We have seen this work, but few people are willing to go to this extent.

If you cannot get them to go away, you may need to trap and relocate them. They are easy to trap the first time but can become trap smart if they have been caught before.

If you decide to live trap and relocate raccoons, keep in mind that raccoons can carry diseases and can be vicious.